Common Waters, Diverging Streams

Linking Institutions to Water Management in Arizona, California, and Colorado

William Blomquist

Edella Schlager

Tanya Heikkila

Routledge
Taylor & Francis Group

NEW YORK AND LONDON

First published by
An RFF Press book
Published by Resources for the Future
1616 P Street NW
Washington, DC 20036–1400
USA
www.rffpress.org

This edition published 2012 by Routledge
711 Third Avenue, New York, NY 10017
2 Park Square, Milton Park, Abingdon, Oxfordshire OX14 4RN

Library of Congress Cataloging-in-Publication Data

Blomquist, William A. (William Andrew), 1957 –
Common waters, diverging streams : linking institutions to water management in Arizona, California, and Colorado / William Blomquist, Edella Schlager, and Tanya Heikkila.
      p.   cm.
    Includes bibliographical references and index.
    ISBN 1-891853-83-X (hardcover: alk. paper)
    1. Water-supply—Arizona—Management.   2. Water-supply—California—Management.   3. Water-supply—Colorado—Management.   4. Water resources development—Government policy—Arizona.   5. Water resources development—Government policy—California.   6. Water resources development—Government policy—Colorado. I. Schlager, Edella, 1960–   II. Heikkila, Tanya.   III. Title.
TD224.A7B58 2004
363.6 '1 '0979—dc22   2004001359

f e d c b a

This book was typeset in Minion by Agnew's, Inc. It was copyedited by Chernow Editorial Services, Inc. The cover was designed by Maggie Powell. Cover photograph by Paul Souders.

ISBN 1-891853-83-X (cloth)   ISBN 1-891853-86-4 (paperback)

About Resources for the Future *and* RFF Press

Resources for the Future (RFF) improves environmental and natural resource policymaking worldwide through independent social science research of the highest caliber. Founded in 1952, RFF pioneered the application of economics as a tool for developing more effective policy about the use and conservation of natural resources. Its scholars continue to employ social science methods to analyze critical issues concerning pollution control, energy policy, land and water use, hazardous waste, climate change, biodiversity, and the environmental challenges of developing countries.

RFF Press supports the mission of RFF by publishing book-length works that present a broad range of approaches to the study of natural resources and the environment. Its authors and editors include RFF staff, researchers from the larger academic and policy communities, and journalists. Audiences for publications by RFF Press include all of the participants in the policymaking process—scholars, the media, advocacy groups, NGOs, professionals in business and government, and the public.

Contents

List of Tables

List of Figures

About the Authors

William Blomquist is associate professor of political science at Indiana University Purdue University Indianapolis (IUPUI). His research interests include the formation of public policy and the management of water resources, with a particular emphasis on the roles and significance of institutions. In addition to his 1992 book, *Dividing the Waters,* his published work has appeared in *Political Research Quarterly, Water International,* and the *Journal of the American Water Resources Association.* He serves on the Research Advisory Board of the National Water Research Institute.

Edella Schlager is associate professor in the School of Public Administration and Policy at the University of Arizona. Her most recent research focuses on the structure and design of public organizations, institutions, and property rights, including water laws and organizations, river compacts, and civil service systems. She has recently participated in a National Resource Council committee on Defining the Best Available Science in Fisheries. She has recently published articles in *Natural Resources Journal* and *American Behavioral Scientist.*

Tanya Heikkila is an assistant professor with Columbia University's MPA program in Environmental Science and Policy in the School of International and Public Affairs. She teaches courses in public management and policy implementation. Heikkila's research interests focus on comparative analyses of public institutions and water resource management. Prior to starting with Columbia University, Heikkila spent a year as a post-doctoral fellow with Indiana University's Workshop in Political Theory and Policy Analysis. She also was the recipient of Resources for the Future's Joseph L. Fisher dissertation fellowship. Heikkila has recently published articles in *Natural Resources Journal* (with Schlager and Blomquist), the *Journal of Policy Analysis and Management, Water Policy,* and the *American Review of Public Administration.*

Preface

Water, a vital yet scarce resource, is the subject of intense competition and great conflict as individuals and groups bargain and struggle to advance and maintain their access to it. It is also the stimulus to considerable creativity and remarkable cooperation as individuals and groups look for ways to preserve and protect a resource on which they jointly depend. Sometimes they must overcome barriers to cooperation, and often they must invent new ways of working together.

The challenges can be even more daunting when the same people are trying to use, share, and protect more than one water source at the same time. This is the case with conjunctive water management, the effort to manage surface water resources and groundwater resources together. Efforts to cooperate and coordinate across individuals, and also across distinct water sources that may be governed by different rules and under the control of different organizations, face even greater obstacles and require even greater creativity. Although these water management challenges can exist anywhere, we have been especially interested by the situation in the western United States, where rapid growth combines with limited and erratic freshwater supplies in ways that constantly press the temptations of competition and conflict against the pursuit of coordination and cooperation.

This book was written with two major purposes in view. One was to contribute to the water resources management literature by describing and documenting conjunctive management in three important states of the American West—how it has come into existence, how it is practiced, what it does and does not accomplish, and how institutional arrangements affect each step in this process. This examination was linked with our other purpose: to contribute to the literature on public policy by connecting specific institutional arrangements to specific policy actions and outcomes in a comparative empirical setting. By comparing three states that face similar water resource problems but have different laws and organizational structures, we have tried to illuminate how institutions matter in the improvement of resource management

The data collected on Arizona, California, and Colorado for this book come from a research project conducted by the authors from 1997 through 2000, with support from the U.S. Environmental Protection Agency and the National Science Foundation. We gratefully acknowledge their support under the NSF/EPA Water and Watersheds Program (Grant No. R824781).

That project benefited from the excellent work of several graduate students at the University of Arizona—Peter Deadman, Nancy Ellis, Todd Ely, Benjamin Hale, and Shalla McMillen. Tanya Heikkila started her collaboration on the project that led to this book at that time, as well.

During that data collection period, we interviewed and collected documents from dozens of individuals in Arizona, California, and Colorado. All of them were generous with their time, and contributed valuable information and insights to our work, and we are grateful to them.

We appreciate the gracious cooperation of the editors at *Natural Resources Journal,* who granted us permission to republish here some of the tables that appeared in our article, "Institutions and Conjunctive Water Management in Three Western States," which appeared in the *Journal* in Summer 2001.

Our thanks also go to the anonymous reviewers of this manuscript. Their suggestions and critiques helped us improve it. Thanks, too, to Don Reisman, publisher at RFF Press, for his good advice about the manuscript and his assistance in shepherding it through to publication. We remain of course solely responsible for its contents and conclusions.

A special word of appreciation is due to Elinor (Lin) Ostrom, for her encouragement during this project and for the many ways in which she has supported and assisted each of us individually over the years. It is our pleasure and privilege to dedicate this book to Lin with gratitude to her as our mentor, colleague, and friend.

Finally, a word of thanks to our loved ones—Kerry, Michael, and Eric, Todd and Claire, and Margaret—whose support is precious to us every day.

W.B.

E.S.

T.H.

Part I

Common Waters: Managing Surface Water and Groundwater Resources Together

1

Water Scarcity, Management, and Institutions

*It is a commonplace to say that institutions matter; it is less
often that one is able to discern precisely in what ways.*
—Gregg et al. 1991, 6

Water has always been vital to the West, inextricably entwined with its development and culture. The allocation, use, and protection of water resources are among the West's most important political and public policy issues. Four factors related to water management now collide regularly and repeatedly in the western United States. The first is the scarcity of water, apparent and well known even to those who live elsewhere. Water scarcity has shaped the region and its states and communities.

Another factor is the variability of water supplies, less well known but just as challenging. The West is not uniformly dry, either in space or in time. Sudden storms can overwhelm channels and banks and levees, making flood losses as great a risk to lives and economies in the region as the more familiar aridity. During the rainy portion of every year, water that later will be sorely missed escapes the region in rivers and runoff. But the areas and periods of abundant rainfall and snowmelt are separated by vast spaces and times of scarcity. Precipitation in the Southwest in particular is concentrated in a few months of the year, followed by long dry seasons. Droughts can even extend across years. Water supplies are always either too plentiful or inadequate, in the wrong location or at the wrong time.

The third factor is the growth of population, agriculture, and industry, producing high and ever rising water demands that persist from one season to the next throughout the year. These combined demands, which have little regard for what season it is or whether the year has been wet or dry, lead to the consumption of available water supplies increasingly often, spurring the need to address the issues of who will get what, when, where, and how. These questions are at the heart of public policy decisions, and in the West, managing water supplies has been and remains an intensely political issue. "Water wars" are perennial features of the western political landscape, especially in the states we focus on in this book—Arizona, California, and Colorado.

Fourth are nature's own water demands, for the region's animal and plant species and the habitat that sustains them. As the demands of economic growth and development have come closer to tapping out the area's water supplies, these environmental needs at times have been left unfulfilled. Increasingly protected by federal and state government regulations, however, the environment itself now presses its claims on the water supplies of this arid locale. Meeting those claims to any significant degree almost certainly means redirecting some water away from current uses—uses that have economic values and organized interests to bring to bear on such decisions.

The collision of these four factors has drawn a great deal of attention and prompted several excellent articles and books on the need for innovation and transformation in western water policy and practice. Two themes that are prominent in that impressive literature are (1) the need to integrate the management of the West's groundwater and surface water resources, a process known as "conjunctive management," and (2) the importance of institutions—laws, policies, and organizational arrangements—in making improved water resource management possible.

This book draws together these two themes and examines the conjunctive management of groundwater and surface water as a means of improving water resource management, as well as how institutions affect conjunctive management. This analysis is based on a study that compares the institutions and water management experiences in the states of Arizona, California, and Colorado. Although the focus is on these three states, the findings in this book also provide insights into the role of institutions in water management generally, within or beyond the American West. Before delving into the three-state comparison, we need to discuss further the status of the western water supply, introduce the concept of conjunctive management for readers who are unfamiliar with it, and set out our ideas about institutions and how they are connected with the adoption and implementation of a strategy for water management improvement such as conjunctive management.

Water Demands and Supplies: Scarcity and Vulnerability

Water scarcity is not a new phenomenon in the American West. The water supply and demand situation at present and for the foreseeable future nonetheless is increasingly difficult. Demands are growing, as are the temporal and spatial imbalances between water demands and supplies. At an increasing number of places and times, there is literally not enough water to meet all needs and claims. Further, the water supplies themselves have grown more vulnerable.

Population Growth Produces Rising Demands, Even with Conservation

Population growth in the West has for some time surpassed that of any other region of the United States. Southwestern states such as Arizona, Colorado, and Nevada have led the nation in percentage increases, and California has added more people than any other state (U.S. Census Bureau 2001). Fifteen of the country's fastest growing counties are in the Southwest. Of these, five had population increases of 90 percent or more—in essence, a doubling—during the 1990s.

Water demand increases are overwhelming the region's impressive gains in water use efficiency. Per capita water use in Arizona, California, and Colorado actually declined in the first half of the 1990s, dropping at a much greater rate than for the United States as a whole (Table 1-1). In 1995, water use per resident in these three southwestern states was nearly 100 gallons of water per day lower than just five years earlier. Despite the continuing success of water conservation efforts, total water withdrawals rose in the three states, and at a much greater rate than for the United States as a whole.

Agricultural Water Use Is Not Declining to Offset This Growth

Table 1-1 implicitly conveys an additional message about rising water demands in the Southwest: increased water demands for the growing population are not being matched by declines for agriculture, long the region's largest water user. Despite many observers' predictions and hopes that reductions in agricultural water use would free up the supplies needed for the West's burgeoning population, it does not appear that irrigation will be giving way soon.

Cities and suburbs displace thousands of acres of farmland per year, and total acreage in farm production in the West declined slightly during the 1990s. Yet the average value of farms and the market value of agricultural products rose, because agricultural production is intensifying on the remaining land. In California, for example, between 1987 and 1997 the number of acres of farmland decreased 18 percent, while the market value of agricultural products sold rose nearly 40 percent (U.S. Department of Agriculture Statistics Service 1997).

Intensification of agricultural production means that reduced agricultural acreage does not necessarily translate into reduced agricultural water demand. For example, water demand for irrigated agriculture in California rose 25 percent from 1960 to 1990, from 20 million acre-feet to 26.8 million acre-feet per year (California Department of Water Resources 1994a). Urban water demand tripled over the same period, but notice the amounts: annual urban water demand grew by 4.5 million acre-feet, from 2.3 to 6.8 million acre-feet (an acre-foot being equivalent to 326,000 gallons) (California Department of Water Re-

Table 1-1. Water Withdrawals and Per Capita Use, 1990 and 1995

	Arizona	California	Colorado	United States
Total freshwater withdrawals (millions of gallons per day), 1990	6,570	35,100	12,700	339,000
Total freshwater withdrawals (millions of gallons per day), 1995	6,820	36,300	13,800	341,000
Percent change, 1990–1995	+3.8%	+3.4%	+8.7%	+0.6%
Per capita freshwater use (gallons per day), 1990	1,790	1,180	3,850	1,340
Per capita freshwater use (gallons per day), 1995	1,620	1,130	3,690	1,280
Percent change, 1990–1995	−9.5%	−4.2%	−4.2%	−0.4%

Source: Solley et al. (1993, 1998).

sources 1994a). Thus, agriculture's 25 percent increase over the 30 years actually involved a larger amount of water than the nearly 200 percent increase for urban residents. Projections to 2020 show a 6.7 percent decline in annual agricultural water use in California compared with 1995, but the reduction of 2.3 million acre-feet will not match the anticipated 3.3 million acre-feet rise in urban water use (California Department of Water Resources 1998a). With population growth outpacing the gains in water use efficiency, and without corresponding decreases in agricultural water use, overall water demands for human consumptive uses continue to rise in the West.

Environmental Water Needs

Tradeoffs among scarce water supplies in the West were long perceived as two-dimensional—the needs of irrigated agriculture versus the needs of growing urban populations. More recently, federal and state agencies and environmental interest groups have been attempting to quantify and guard a third category of water needs to sustain riparian or aquatic species and habitat.

On the one hand these are not new water needs—the species and habitat of the West have depended on the water supplies of streams and lakes all along. On the other hand, recognizing environmental water needs intensifies the water scarcity situation in the West. To date, the rising needs of people and crops have been met by drawing down the environmental water account—that is, by reducing the flows in streams or the levels in lakes and reservoirs when needed, regardless of the importance of those flows or levels to species and habitat. Taking environmental water needs explicitly into consideration dispels the illusion that "more water" exists to be tapped for people and crops.

This effect is perhaps most visible in California's state water plan, which is updated every five years by the California Department of Water Resources. Beginning in the early 1990s, California Water Plan updates included estimates and projections of environmental water requirements along with urban and agricultural needs. The addition of environmental needs made it clear that there simply was no "more water" for farms and cities to obtain. In fact, the combined urban, agricultural, and environmental needs in a typical year *already* exceed the state's average annual water supplies by one million acre-feet or more, with the deficit being sustained primarily through groundwater depletion (California Department of Water Resources 1998a). Such figures have clarified that future increases in water demands for people or crops will have to be met in some other way; otherwise the fulfillment of these needs will continue to come at the expense of the environment.

In other cases, environmental water needs are being added to water supply systems that are already fully appropriated for urban and agricultural uses. In 1994, the Arizona legislature created the Arizona Water Protection Fund, which supports projects that protect, maintain, or enhance Arizona's rivers and riparian habitat (Ariz. Rev. Stat., Title 45, Section 12). The Central Arizona Project plays a central role in such projects as both a source of funding and water. The Platte River system, which includes Wyoming, Colorado, and Nebraska, also provides an acute example. The three states have sparred for decades over the waters of the Platte system. The South Platte in Colorado has long been determined to be fully appropriated, that is, existing recognized water rights along the river equal or exceed the river's average annual flow. But the middle reach of the Platte River in Nebraska is home to federally protected endangered species and habitats, and the U.S. Fish and Wildlife Service has indicated a need for an additional 417,000 acre-feet per year of water in the system to support and sustain them. Procuring hundreds of thousands of acre-feet of additional water in the Platte system means nothing less than reducing other established water uses. This is water scarcity in its starkest manifestation: more for one category of use means less for the others. The effects of this gridlock over Platte River supplies spread out in several directions:

> Endangered species requirements, as defined by the USFWS, essentially preclude any water consuming action that constitutes a federal nexus. This means that U.S. Forest Service leases in Colorado cannot be renewed; that Wyoming cannot pursue additional upstream water storage projects that would increase consumptive use; and that the public power districts in Nebraska cannot get a long-term hydropower license from [the Federal Energy Regulatory Commission] unless some accommodation of the competing demands can be made. (Supalla et al. 2002)

Meeting environmental water needs under such circumstances has also meant changing the operations of surface water projects that were originally built to supply the other uses. Water releases from dams, and the maintenance of water flow and quality levels in canals and aqueducts, are to an increasing degree based on the needs of aquatic species dependent on stream flows. Under the Central Valley Project Improvement Act (CVPIA), Section 3406(d)(2), for example, the U.S. Bureau of Reclamation has to acquire water for delivery to wetland habitat areas in the Sacramento and San Joaquin valleys in California. The CVPIA obligates the Bureau to assure 80 percent of water needs for the Kern National Wildlife Refuge in the 2000–2001 water year, 90 percent in 2001–2002, and 100 percent in 2002–2003 and beyond. The Bureau has purchased water for the refuge from local California water agencies with stored surplus underground water available for sale.

Other Instream Flow Needs

Riparian and aquatic species and habitat are not the only claimants for river and stream flows. Hydropower generation has been an important energy source for the West for a century, and holders of federal hydropower licenses continue to need adequate flows to maintain production at their facilities. Although hydropower has declined as a share of total electricity production in the West, the electricity problems of California and the Pacific Northwest in 2001 made it clear that all power sources will continue to be needed for the foreseeable future.

The changing economy of the western U.S. has increased the value of instream flows for other uses as well—commercial and recreational fishing, recreational boating and rafting, and other aspects of the emerging "ecotourism" industry. Colorado water law now includes a category of "recreational in-channel" water rights, defined as a minimum stream flow between two points for reasonable recreational purposes, and several communities have pursued such rights in order to satisfy residents' and visitors' desire to enjoy kayaking or other stream-based activities. The prominence of recreational industries is evident in the debate over the decommissioning of dams on the Snake and Columbia rivers, where environmental groups and commercial fishing interests have teamed up to argue that the economic value of the recreational and commercial uses of the river flows, especially the salmon stocks the rivers support, now exceeds the economic value of the electricity generated with the dams.

Conflict over river flows and the operation of dams and reservoirs on America's longest river, the Missouri, has sharpened considerably during the past decade. For decades, the predominant economic activity on the Missouri was barge transportation, which is concentrated in the lower reaches of the river above its confluence with the Mississippi. Up-river dams and reservoirs were operated by the U.S. Army Corps of Engineers not only for downstream flood protection, but also to manage river levels for the benefit of the shipping industry, which remains economically important to the down-river states of Nebraska and Missouri. Lately, however, those reservoirs in the up-river states of South Dakota, North Dakota, and Montana have become economic centers of their own, hosting a recreation and fishing industry worth $85 million per year to the region. That recreation business depends on maintaining relatively consistent and high reservoir levels, while the down-river shipping industry has depended on releases from those reservoirs during dry periods. As in many areas of the West, the Missouri River simply cannot supply enough water to meet all demands at all times. The Corps has been taken to court at both ends of the river, ordered by judges at one end to halt further releases that jeopardize reservoir levels and by judges at the other end to resume releases immediately to maintain river flows (Pianen 2002).

Instream uses compete not only with one another—fish versus dams, recreational lakes versus river transportation—but also with out-of-channel withdrawals of water for use on the land. Especially during dry times of the year and during droughts, water removed from streams and lakes for irrigation or municipal uses is unavailable for those instream uses—the situation is essentially zero-sum. Last but not least in regard to instream uses, as a growing number of rivers in the West acquire "wild and scenic river" designations, their flows receive additional protection under federal and/or state laws.

Time, Space, and Vulnerability

Western water problems do not consist solely of imbalances between annual supply and annual demand or of conflicting demands for scarce resources. Issues pertaining to time and space aggravate the underlying problem of scarcity.

First, large urban populations and high-intensity agricultural production require water throughout the year, but precipitation and stream flows are not steady from month to month. In the Southwest especially, these natural water supply sources are concentrated in the winter and early spring.[1] Peak water demands occur in the summer and fall, for irrigation, recreation, drinking water, and waste disposal. Hydropower interests also would prefer maximum reservoir releases in the summer to generate enough flow to produce electricity when demand and prices are highest.

Second, western settlement and development patterns have often located people and agriculture away from the most plentiful water sources. California and Colorado are among the most glaring examples. The majority of water supply—precipitation, snowpack, and stream flow—is found in the northern half of California and the western half of Colorado, but the majority of agricultural and urban water demand is located in the southern half of California and the eastern half of Colorado. In these two states, as elsewhere in the West, water has been moved from places of comparative abundance to places of greater need. California's aqueducts and Colorado's transmountain diversions are legendary, but they are not unique. As the native water supplies available to Phoenix and Tucson became insufficient to meet their rising demands, Arizona and the federal government undertook the Central Arizona Project, which diverts water hundreds of miles across the desert to those metropolitan centers.

The distances between growing centers of water demand and the natural sources of water supply are lengthening. In Southern California, for example, development accelerated during the 1990s in southern Orange County and northern San Diego County—areas far from sizable rivers and not overlying substantial groundwater deposits. These communities are almost entirely dependent on pipelines conveying water to them from distant sources.

Combined, these time and space discrepancies between water demands and supplies accentuate the vulnerability of water supply and distribution systems. Early in the twenty-first century, water reliability is becoming as important in the Southwest as water availability. Aqueducts and pipelines that cross dozens or even hundreds of miles are exposed to natural, accidental, or even deliberate disruption or destruction. Reservoirs and centralized water treatment plants face similar risks.

The public health and economic consequences of supply interruptions, or even supply diminutions, are enormous. An economic impact analysis was conducted in 1993 and updated in 1999 for the San Diego County Water Authority. The figures, for just part of one county in the contemporary American West, were remarkable. The study estimated that a two-month reduction of just 20 percent of water supplies would result in combined employment and other economic losses of $2.3 billion, rising to $13 billion if as much as 60 percent of water supply were unavailable for the same period. Six-month interruptions carried estimated impacts ranging from $8 billion for a loss of 20 percent of supplies to $32 billion for a loss of 60 percent (CIC Research 1999).

Vulnerability to contamination is another concern. The concentration of the region's growing population in urban and urbanizing areas, bringing together large numbers of people and businesses and the wastes they generate, increases opportunities for water supply contamination. Intensified agricultural production can have some of the same effects, from large-scale livestock operations to the fertilizer and pesticide runoff of intensively farmed cropland.

Each of these phenomena—population growth, urbanization, agricultural intensification, peak water demands that are out of phase with the timing of water supplies, and lengthening distances between those supplies and the areas of growing demand—contributes to concerns about water supplies, water storage, and water system reliability. Combined, their effects present great challenges to the West's water future. In their report, *Envisioning the Agenda for Water Resources Research in the Twenty-First Century,* the members of the National Research Council's Water Science and Technology Board emphasized that challenge: "The progressive intensification of water scarcity in the early decades of the twenty-first century will necessitate innovative scientific, technological, and institutional solutions" (National Research Council 2001, 45).

Water Storage and Conjunctive Management

Ideas for solutions are not scarce. Several are being advocated and pursued vigorously throughout the West. On the water-demand side, they include further conservation initiatives, water-use efficiency requirements or incentives, growth restrictions, water marketing, and water-pricing reforms. On the supply side they include further development of water purification technologies for desalination and wastewater recovery, to develop or reclaim water sources that have heretofore been unusable.

Water Storage Needs in a New Political and Economic Environment

Even together, however, these important and useful solutions will likely fall short of addressing the water scarcity and vulnerability problems currently at work in the West. What communities throughout the contemporary West need, and will continue to need, are buffers. In particular, they need buffers against

- peak demands that exceed economically feasible supply system capacities;
- seasonal variations in surface supplies that make summer and fall demands difficult to sustain;
- droughts of varying intensities and durations;
- contamination of a water source or distribution system;
- accidental or deliberate interruption of surface or imported supplies and facilities; and
- endangerment of stream flow for species, habitat, and downstream communities.

Such buffers are provided by water *storage,* not just by reducing total demand or increasing the flow of supply. The volume, and more importantly the time-and-space pattern, of modern water demands in the West cannot be met reliably without extensive storage and distribution facilities. Matching supplies that arrive in one part of the year with demands that peak in another, or water demand that occurs in one location with water supplies that arrive in another, requires large amounts of water storage capacity.

Throughout most of the twentieth century, needs for extensive water storage capacity and distribution facilities were met by constructing dams and reservoirs, canals and aqueducts, and by diverting water from streams and across the land to points of need and use. Today, responses to concerns about water supply, storage, and reliability take place in a changed political, economic, and cultural environment. Above-ground storage capacity is more difficult to construct at present because of the diminished number of available sites, the expense of land acquisition and facility construction, the environmental damage of further disruptions to streams and their natural channels, and the West's changing economy and culture.

Changing water demands and values mean that the rising problems of water scarcity, storage, and reliability in the Southwest are likely to require and elicit different policy responses (National Research Council 2001, *13*). Conjunctive water management has been one of those responses. Here we introduce the concept of conjunctive management briefly, before moving on to the institutional issues and the questions we want to address in this book.

Conjunctive Management

The human-built surface water projects of the contemporary Southwest are the visible, dramatic, and often described components of the region's water resources. But resting quietly beneath the surface, the region's groundwater assets are larger than even the best known concrete structures above the ground. California's groundwater basins, for example, hold an estimated 850 million acre-feet of water—approximately 20 times the quantity that can be stored behind all of the state's dams combined (Association of Ground Water Agencies 2000, *4*).

Throughout the Southwest, groundwater supplies have been tapped to provide a source of relatively high-quality, comparatively inexpensive water close to points of use. In all three states that are the focus of this book, groundwater has supported the growth of irrigated agriculture *and* large urban populations, especially after high-capacity pumps were developed in the early decades of the twentieth century. Groundwater supplies in all three states have even been overdrafted—that is, withdrawn at a greater rate than they are replenished by the percolation of precipitation and stream flows into the ground. Indeed, while the prospects for increased surface water storage capacity in the West have dimmed, the overdrafting of groundwater supplies and the lowering of water tables has *increased* the amount of available underground water storage capacity.[2]

Conjunctive water management involves the coordinated use of surface water supplies and storage with groundwater supplies and storage. At times when surface water supplies are comparatively plentiful—in winter months and during wet years—conjunctive water management encourages the direct use of those surface supplies as well as their storage to the extent possible behind dams or other impoundment structures. Groundwater basins are tapped less during these wet periods, allowing them to refill naturally or through deliberate replenishment efforts. When surface water supplies are comparatively scarce and may need to be conserved for instream flow needs, stored groundwater can be tapped to supply irrigation or urban uses. Thus, in a given location, the same water uses may be met at some times with surface water and at others with groundwater, as part of a deliberate management effort that operates the two resources conjunctively (U.S. Advisory Commission on Intergovernmental Relations 1991). To some extent, conjunctive use of surface and groundwater supplies results from water users' actions even in the absence of deliberate management, as users with access to more than one water source switch back and forth according to relative availability.[3]

Conjunctive management is not a new idea. It has been practiced in some southwestern communities for three-quarters of a century (Todd and Priestaf 1997, *139*). It has been advocated in the professional literature on water resources management for more than 50 years (Conkling 1946; see also Banks 1953; Thomas 1955). Conjunctive management has drawn greater attention in the West recently, however, as the region's water scarcity, storage, and reliability concerns have escalated and the ability to deal with those concerns through conventional water development methods, such as the construction of new dams and reservoirs, has waned.

Conjunctive management can be seen as a method of addressing these resource management dilemmas that appear throughout the region. It offers the promise of greater average annual water supply yields by capturing and conserving surplus water supplies when they are available, reducing losses that would occur if surface water or groundwater sources alone were relied on. It also can allow some surface water supplies to remain in streams for environmental and recreational purposes without a zero-sum reduction in the amount of water available for human consumptive uses. Furthermore, conjunctive management serves these purposes with less of the expense and environmental damage associated with the construction and operation of additional surface water structures on the scale of the Southwest's current and projected future needs.[4]

Because of its inherent qualities and its relative advantages compared with other water supply alternatives, conjunctive management has been one of the more popular recommendations for improving the water resource situation in the West generally and in the Southwest in particular. The Natural Heritage Institute heralded conjunctive management as a form of "environmentally benign water development," and recommended "using groundwater storage to make sure that both the environment and the economy have the water they need in dry years" (Natural Heritage Institute, 1997, *2;* see also Long's Peak Working Group 1994). Based on this potential, state and local water agencies have been exploring, operating, or expanding conjunctive management throughout the region, as our analysis of California, Arizona, and Colorado demonstrates. Given the disparity between water supply and demand situations in many states, it is clear that a large number of communities have yet to develop policies and tools for resolving water resource dilemmas, especially the vexing problems of the western United States.

Connecting Institutions to Water Management

What then explains why some communities choose to implement more progressive water management programs, such as integrating the management of ground and surface water supplies, while others have not? In the case of conjunctive water management, the technical aspects are straightforward enough. One captures, conserves, and distributes surface water supplies when they are available, and stores them underground when they are in surplus. One supplements surface water supplies with groundwater as needed to get through peak demand periods. In very dry periods, when surface water is unavailable or devoted entirely to instream uses, one may switch over to groundwater altogether. Why would the implementation of such a straightforward and sensible management approach continue to lag behind its promise, even after decades of support in the water resource management literature?

A recent experience offers a clue. At a May 1998 Workshop on Conjunctive Use convened by the National Water Research Institute and sponsored by the Association of Ground Water Agencies and the Metropolitan Water District of Southern California, participants from local, regional, and state agencies as well as the academic and consulting communities identified and ranked "impediments to implementing a cost-effective conjunctive use water management program in California." In order of priority as ranked by the participants, the 10 most significant impediments were (National Water Resources Institute 1998, 5–39)

- an inability of local and regional water management governance entities to build trust, resolve differences both internally and externally, and share control;
- an inability to match benefits and funding burdens in ways that are acceptable to all parties, including third parties;
- a lack of sufficient federal, state, and regional financial incentives to encourage groundwater conjunctive use to meet statewide water needs;
- legal constraints regarding water rights, groundwater basin management, authority to store water underground, and authority to transport water across basin boundaries that impede conjunctive use;
- a lack of statewide leadership in the planning and development of conjunctive use programs as part of comprehensive water resources plans, while recognizing local, regional, and other stakeholders' interests;
- inability to address differences in quality among different types of water and unforeseeable future groundwater degradation;
- risks that water stored cannot be extracted when needed because of infrastructure, water quality or level, politics, and/or institutional/contractual provisions;
- a lack of assurances to address third-party impacts and to increase willing local citizen participation in conjunctive use projects;

- a lack of creativity in developing lasting "win–win" conjunctive use projects, agreements, and programs; and
- supplemental suppliers and basin managers have different roles and expectations in relation to conjunctive use.

A majority of the workshop participants were engineers, hydrogeologists, and water agency managers. Despite the participants' professional and occupational backgrounds, *none* of the impediments that made the group's "top 10" list were physical or financial per se. Every one of the 10 most significant barriers they identified concerned the assignment of rights, risks, and responsibilities; the distribution of costs and benefits; and the opportunities and disincentives for interorganizational cooperation and coordination of activities—in short, institutional issues.

As the "top 10" list of impediments implies, conjunctive management requires a large amount of joint and coordinated effort among individuals and organizations; and institutions are key to coordination (Runge 1984).[5] Even where conjunctive management is physically possible and economically feasible, whether and how it is pursued depends on the institutional setting. The coordinated actions necessary for implementation of a conjunctive management program are more likely to occur if institutional arrangements (1) promote the actions necessary to divert, impound, recharge, store, protect, and extract water or (2) protect those who invest in facilities, or who store water now for recapture later, and (3) provide or recognize workable and fair methods for distributing the costs of a conjunctive management program among those who benefit from it.

It is nothing new to make the argument that institutions matter in water resource management (Ingram et al. 1984; Livingston 1993; Lord 1984). However, references to institutions in the literature on water resource management often fall into one of three categories—institutions as black box, institutions as scapegoat, and institutions as deus ex machina—without fully explaining *how* institutions matter.

- "Black box" references treat institutions or institutional factors as a shorthand for politics, government, legislative or regulatory policies, court decisions, and so forth. To some extent, what occurs in these domains and processes is recognized as important, but little appetite exists to peer inside that "black box" and try to understand and explain what happens there—understandably, as most water policy writers are not political scientists.
- "Scapegoat" references treat institutions and institutional factors as a sort of catchall reason why otherwise perfectly sensible, efficient, fair policy prescriptions have not been adopted. Thus, we see contributors to the water management literature touting the advantages of water transfers, reuse, conjunctive management, or some other option as the best answer to a partic-

ular water resource problem, then faulting "institutional factors" as the obstacle impeding its implementation.

- Deus ex machina references call on institutional change to save or rescue a situation. The water policy literature is filled with appeals to institutional change as the solution to a water management problem—transform public to private ownership, change from regulatory to market-based allocation, centralized water policy, or decentralized water management. These are like recommending "flipping the institutional switch" to achieve desired policy goals.

We believe, therefore, that much remains to be gained by a close and thorough consideration of how institutions are connected with water resource management—how they stimulate changes in management practices, how they facilitate or hinder those changes, how they shape the management choices water users and organizations make, and how they affect the outcomes water users and organizations achieve. A comparative institutional analysis seemed to us a sound way to pursue those questions.[6]

We designed this comparative study to try to improve academic as well as the general public's understanding of the effects of institutional arrangements governing the allocation, use, and protection of water resources on the development, implementation, and performance of conjunctive management programs. Comparative institutional analysis is challenging because of the intensive data needs and the large and uncontrollable number of other confounding factors that are inevitably associated with comparing jurisdictions that have diverse histories, populations, economies, and so forth. Empirical studies of the relationship between institutional differences among the states and observed outcomes in the development and implementation of conjunctive management programs are unusual, to say the least. In the particular context of applying game theory to the study of groundwater management, Gardner et al. (1997, *219fn*) observed, "The costliness of collecting data on groundwater use and the difficulty of applying game-theoretic models explains the overwhelming reliance . . . on analytical results, simulation methods, or reasoned institutional arguments concerning the desirable properties of groundwater property-rights systems." They are of course correct, and we have great respect for modeling, simulation, and other analytical methods of approaching these issues. An important role also exists for empirical studies, despite their costliness in time and monetary expense.

Comparing Arizona, California, and Colorado

To advance the understanding of institutions and their effects, and to do so in a comparative empirical context, it is essential to find cases that differ on crucial institutional variables of interest but that are reasonably similar in other respects. Although conjunctive management is practiced in a variety of locations throughout the United States, we selected Arizona, California, and Colorado for this study because of their similarities and differences on key variables of interest to us.

Arizona, California, and Colorado share several relevant characteristics—their rapid population growth and rising water demands, great discrepancies between the geographic distribution of water supplies and water demands, location in the arid Southwest, exposure to lengthy and severe dry periods, and employment of conjunctive management, at least in parts of each state. The three states differ substantially, however, in their institutional arrangements governing and managing water supplies, including their water rights laws and the types and authority of water management organizations found in each state.[7] We were particularly intrigued by (1) Arizona's relatively new (post-1980) statewide policies governing groundwater management, (2) Colorado's watershed-level system of water governance and management, and (3) California's decentralized system of local special districts involved in conjunctive management. The combination of physical similarities and institutional differences presented an opportunity to study more closely and attempt to identify the ways in which the institutional differences appeared to be linked to differences in the purpose, organization, operation, and performance of conjunctive management programs in the three states.

Similarities of the States' Water Supply and Demand Situations

As in other arid western states, native surface water supplies are limited in Arizona, California and Colorado. Average annual rainfall in the most populated regions of each state ranges from 8–12 inches in Arizona to 12–20 inches in southern California and eastern Colorado (Western Regional Climate Center 2001).

All three states face significant growth pressures that threaten these scarce water supplies. All three states experienced large population gains from 1990 to 2000 (Table 1-2). Arizona was the second fastest growing state in the nation that decade, and the Phoenix–Mesa metropolitan area in Arizona was the eighth fastest growing in the country. California added more than four million residents, and some Southern California counties grew by 25 percent or more. Colorado contained eight of the fastest growing counties in the United States, each of which grew by more than 80 percent during the decade.

Despite the states' large and rapidly increasing populations, water withdrawals for municipal and industrial uses in Arizona, California, and Colorado are still minor compared to those for agriculture. In each state, agriculture accounts for approximately 75–80 percent of total water use (U.S. Department of Agriculture 2001).[8] Municipal, industrial, and recreational users face increasing competition for scarce water supplies that are directed largely to irrigation.

Table 1-2. Population Change in the Three States, 1990–2000

	1990 population	*2000 population*	*Percentage change*
Arizona	3,670,000	5,130,000	40.0
California	29,810,000	33,870,000	13.8
Colorado	3,290,000	4,300,000	30.6

Source: U.S. Bureau of the Census.

All three states have extensive groundwater resources. As with surface water, however, groundwater supplies have been strained in each state. In parts of Arizona, groundwater overdraft has led to land subsidence, diminished well yields, degradation of water quality, and the drying up of streams and rivers (Arizona Department of Water Resources 1999a–c; 2001b). Groundwater overdraft is a similarly serious problem in many of California's coastal and valley basins. In Colorado, increased groundwater pumping has drawn water away from associated surface streams, producing decades of conflict between surface water rights holders and groundwater pumpers.

As mentioned earlier, all three states experience great water supply-demand maldistribution. Precipitation and stream flow are concentrated in the winter months, whereas demand is greatest in the summer. Also, the areas of greatest water demand in each of these states are located far from their renewable surface water supplies. In all three states, local, state, and federal agencies have financed and built major projects to move water to the areas of greatest demand. The projects of greatest significance for conjunctive management are described in the chapters about the states later in the book.

Differences Among the States in Water Management and Institutions

Despite the states' similarities with rising demands in arid environments, the management practices of the states are quite different. This includes their approaches to conjunctive water management. The diversity of the conjunctive management projects in the three states adds to their usefulness for comparative analysis. Some have operated for decades, while others are quite recent. They range in scale from small single-basin recharge projects to large interwatershed transfer and storage operations (Western States Water Council 1990; U.S. Bureau of Reclamation 2000).

The institutional arrangements for managing water resources differ among the states as well, which the chapters that follow will consider. One of the differences in water management institutions in Arizona, California, and Colorado is water rights. As we discuss more specifically in Chapter 3, water rights affect the costs of engaging in conjunctive management. Each state has distinct constitutional and legislative rules defining rights and duties of individuals regarding the capture, allocation, and use of both surface water and groundwater. The Arizona legislature, for instance, implemented policies in the 1980s and 1990s quantifying rights to use groundwater supplies and recognizing rights to store water underground. California and Colorado, on the other hand, combine state laws and permit processes with locally defined rules or court adjudications to determine water rights.

The states also are distinct in the extent to which they integrate pumping rights with surface water diversion rights.[9] Colorado has integrated rights to pump tributary groundwater with its surface water rights doctrine, whereas California and Arizona generally have distinct legal doctrines governing ground and surface waters. How these different water rights doctrines positively or negatively influence the successful implementation of conjunctive water management is a key point of our empirical analyses.

Arizona, California, and Colorado also rely on different organizational arrangements for managing water resources. Each state maintains unique systems of private and public organizations, agencies, and water districts that manage water resources, administer laws, and regulate water use. In general, Arizona's system of water administration is relatively centralized, guided by state agencies. California, on the other hand, maintains a highly decentralized system of groundwater management and administration, relying heavily on local special districts. Colorado relies on regional water divisions for water governance and management, which correspond with the state's major watersheds.

Differences such as these in water management organizations across Arizona, California, and Colorado are tied to the larger institutional settings of these states.[10] Arizona, for instance, tends to have a relatively centralized system of governance overall. Local governments have comparatively little autonomy in Arizona. This characteristic is reflected in the fact that most of Arizona's water management policies have come from the state legislature and state agencies.

Colorado and California have more decentralized systems of governance. These states rely on local institutions and organizations to address and resolve problems, and those traditions are evident in the conjunctive management experiences of the states as well. Particularly with respect to the management of groundwater, state government in California plays almost no role at all, leaving the issue to local governments and the courts (Blomquist 1992).

The organizations developed to manage water resources in Colorado differ from those in California. Colorado state government is involved in administering water policies to a greater degree than California's, but Colorado organizes its administration according to the watershed divisions. The state's activities focus more on enabling and supporting local level governance than they do on direct regulation and management. Water users in Colorado have a long tradition of common-law governance and watershed-level organization of resource management (Schlager 1999).

Because water supply situations in the three states are similar, but their institutional arrangements and their water management practices differ, a comparative institutional analysis presents an opportunity to connect the institutional differences among the states with their distinct management practices. Making those connections explicit is the task of the reminder of the book.

A Look Ahead

Certainly conjunctive management has been organized and conducted differently in different places for reasons relating to institutional arrangements. Conjunctive management projects have performed as intended and over long periods in some locations for reasons that can be attributed to the institutional arrangements water users crafted and put in place. At the same time conjunctive management has fallen short of people's hopes for it nearly everywhere, also for reasons that relate to institutional arrangements.

An explanation based on institutional analysis of how differences among the states might be linked to differences in conjunctive management practices and performance led us into the study and is described in Chapter 3. In the course of the study we gained an additional perspective on how institutions matter. This book is our attempt to share not only the findings with respect to the questions we posed at the outset of the study, but also what we learned along the way.

Of particular significance is the way in which institutional arrangements in each state shaped the very reasons for which conjunctive management is undertaken there, in addition to affecting the extent and the manner in which it is carried out. We began with the assumption that "improved water resource management" was the motivation or purpose for which people undertook conjunctive management, in these three states or anywhere else. Well into our study, we came to understand that the *purposes* for which people in the three states were undertaking conjunctive management varied substantially too— that the purpose was not just a general pursuit of improved water management. Instead, institutional arrangements in each state influenced why water users and organizations in that state were pursuing conjunctive management in the first place.[11]

Once water users or organizations chose to pursue conjunctive management, those institutional arrangements plus others determined why and how they implemented conjunctive management in the manner they did, and the effects that resulted. Institutional factors have also shaped the means by which conjunctive management is performed, that is, the types of projects found in the three states. The purposes of conjunctive management and the means by which it is conducted have, in turn, led water users and public officials to develop certain organizational forms and other institutional arrangements that are distinctive in each state. This is, in short, how we have seen institutions matter in California, Arizona, and Colorado, and what we present and explain in the remainder of the book.

The remainder of Part 1 is devoted to conjunctive management and institutional analysis. Chapter 2 explains why and how conjunctive management is performed, and how it relates to the changing water resource situations and policy environment in the West. Chapter 3 presents our theoretical approach, setting out how institutional arrangements and other factors may influence the adoption and implementation of water management practices such as conjunctive management.

In Part 2, we describe the relationship between water management institutions and conjunctive management activities, in chapters covering California, Arizona, and Colorado, respectively. Part 3 contains comparative analysis of the three states, conclusions about the effects of institutions on conjunctive management, a look ahead to future developments, and some policy recommendations.

2

The Promise of Conjunctive Water Management

Conjunctive management is one of the most popular contemporary recommendations for improving water resource use and protection. As noted in Chapter 1, it involves coordinating surface water supplies and storage capacity with groundwater supplies and the storage capacity of underground aquifers. In this chapter, we discuss conjunctive management more fully, including the purposes it is supposed to fulfill, the methods by which it is accomplished, and recent trends that have contributed to its popularity as a policy recommendation.

Why It's Done: Goals of Conjunctive Management

As we discuss throughout the book, a specific conjunctive management project in any particular location is intended to serve time-and-place–specific purposes. Nevertheless, it is possible to describe at a general or conceptual level the water management purposes served by conjunctive use. Conjunctive management may be understood as an effort to use the relative advantages of surface water and groundwater resources to offset each other's shortcomings. In more practical terms, conjunctive management is intended to reduce exposure to drought and flooding, maximize water availability, improve the efficiency of water distribution, protect water quality, and sustain ecological needs and aesthetic and recreational values.

The Conjunctive Management Idea

Water on the surface of the land—lakes and seas, streams and rivers, and the precipitation and runoff that feed them—provides several obvious values. It is of course a source of life-sustaining drinking water for humans and other land-dwelling creatures. It is home and habitat for aquatic and riparian plants and animals. Its flows can be converted into a useful and renewable energy source. Surface water resources are also valued for the beauty they add to the landscape and the recreational amenities they provide.

Surface water resources also have their shortcomings, from the standpoint of managing the resource for human and other uses or needs. Surface water availability frequently depends on precipitation, which is variable across space and time in both predictable and unpredictable ways. Surface water can be too abundant in one place and inadequate in another, or too abundant at one time and inadequate at another. Through the construction of dams, tanks, canals, and other devices, people have been trying for millennia to overcome this variability of surface water resources.

Uncontrolled, excessive surface water can be destructive. On the other hand, if not captured when it is abundant, surface water slips away to the ocean or some other sink from which it cannot be recovered later for consumptive use if needed. Even successfully stored surface water must be transported if it is to serve any offstream uses; overland transportation of water by aqueducts or pipes or other means is expensive.

Impounded or flowing freely, surface water evaporates—and at the greatest rates in the most arid locations. It is also exposed to direct contact with contaminants from the land or air. These contaminants may originate from natural or human-created sources, but in either case they degrade the value of surface waters for other uses.

Underground water resources—aquifers and the water they contain— exhibit a different combination of strengths and vulnerabilities. Of course, groundwater is variable across space and time, too, but not in exactly the same ways as surface water. The spatial variation of groundwater availability is not just the same as that of surface water. There are places with plenty of both and places without enough of either, but there are also places that are rich in groundwater and poor in surface water supplies or vice versa.

Like spatial variation, temporal variation in groundwater follows a different pattern from that of surface water. Precipitation and runoff are ultimately the sources of groundwater as they are of surface supplies, but accumulating and moving through underground soils shifts the ebb and flow of groundwater supplies to a different, slower cycle. Groundwater may remain plentiful in an area for some years after a drought has taken its toll on water supplies at the surface. A high underground water table can therefore sustain some base flow in streams or some soil moisture in wetlands even when surface supplies have grown scarce or inadequate to do so (National Research Council, 1997, *39*). By the same token, diminished underground water levels take longer to recover even when precipitation and runoff inundate the land surface above.

Nature has provided many times more underground storage capacity than the surface water storage facilities humans have constructed, even after being at the task for thousands of years. Certainly in the southwestern United States, the focus of this book, the amount of water stored and the volume of water storage capacity that exists beneath the surface of the earth dwarfs the most gigantic reservoirs Americans have built above ground (Association of Ground Water Agencies 2000, 3).

In addition to storage capacity, groundwater supplies offer other advantages. First, water stored underground in an aquifer does not evaporate, nor does it flow rapidly downstream or out to the ocean. Of course, groundwater does move from higher to lower elevations or pressures, but these movements are substantially slower than those occurring at the surface. This means groundwater supplies tend to be more secure in the short term, especially in drier climes. Second, groundwater basins and the aquifers they contain[1] generally do not follow the same boundaries as the channels and beds of surface water bodies. This distinction can provide advantages from the standpoint of human uses. Whereas surface water must be diverted and transported to serve purposes away from the channel or bed, it may be possible to extract groundwater with wells located closer to the desired point of use. Third, compared with surface water, groundwater is not as readily exposed to contaminants from the land and the air. Movement of water through soils even provides some natural filtration, intercepting some types and amounts of contaminants (National Research Council 1997, 36).

In addition to these several advantages, underground water resources have significant limitations that affect water users and managers. Diminished groundwater levels can result from overdrafting an aquifer. Groundwater overdraft can produce a host of undesirable effects. The simplest and most temporary effect is financial—more energy is required to lift groundwater to the surface when the water has to be raised from greater depths, and increased energy use raises the cost of water to the user. In locations where aquifers interact closely with overlying streambeds, groundwater overdraft can also dry up surface water supplies, with associated losses to human uses as well as riparian and aquatic species (Kondolf 1994, 135; National Research Council 1997, 37). Other serious, permanent damage from overdrafting groundwater comes in the form of soil compaction and land subsidence, which can degrade or destroy buildings, transportation arteries, and utilities such as water and sewer lines.[2]

Groundwater is also vulnerable to quality degradation. Overdrafting can cause excess minerals to leach into aquifers from surrounding soils, or draw deposits or plumes of contamination toward wells. Overdrafting in a coastal aquifer may reduce an aquifer's water level or degree of water pressure, allowing salt water from the ocean to intrude into the fresh water supply. Although water quality threats such as these may reach groundwater supplies more slowly or in smaller amounts than they do surface supplies, they are less easily flushed out of groundwater once they do invade.

Although groundwater may provide base flow for a stream or wetland, it does not serve as a habitat for aquatic or riparian species in the conventional sense of the word. It certainly does not provide the aesthetic and recreational opportunities for which people value surface water resources.

The fundamental concept of conjunctive water management is to look beyond these contrasts between surface and underground water resources and find complementarities. Through coordinated use, those complementarities can be made to serve practical purposes for human and other uses and needs.

Practical Goals of Conjunctive Management

By coordinating the operation of surface water facilities and the groundwater resource, people can store and protect greater quantities of water over longer periods of time. This coordinated use—conjunctive management—can allow the stored water to provide a wider range of values, from consumptive uses such as drinking to ecological benefits such as conserved or augmented stream flows (National Research Council 1997, *21*).

Among the most obvious benefits people seek to achieve through conjunctive management is reduced exposure to drought and flooding. Above-ground water storage capability, such as dams and reservoirs and enclosed tanks, can be useful for flood control and to retain current surface water for future needs, particularly where extreme temporal variations in surface flows exist. In combination with surface water storage, storing water underground enhances the prospects for withstanding future shortages. As surface supplies dwindle, groundwater reserves can be drawn upon (Knapp and Olson 1995). In the meantime, the stored groundwater has been less vulnerable to losses from evaporation and contamination. By reducing losses, the coordinated use of surface and underground water resources can actually increase the quantity of water available in a particular location over time.

Storing "surplus" surface water, such as unallocated water and storm flows, underground can also reduce groundwater overdrafting and its harmful effects. Drawing down groundwater supplies during periods of diminished surface water makes sense, but is sustainable over the long term only if the groundwater supply is restored when surface water is available. Conjunctive management is one way to make this restoration of groundwater levels deliberate, rather than leaving it to chance.

Moving surplus surface water supplies to underground aquifers also aids flood control. During an especially wet season, a significant danger arises from overfilling of surface water reservoirs. A full reservoir lacks capacity with which to capture the next flood flows that may come along. Recharging surface water to the underground, or encouraging water users to take more surface water in lieu of groundwater, provides opportunities for getting more water out of surface reservoirs during wet periods.

Reducing exposure to drought and flooding are crucial practical objectives of conjunctive management. Several other advantages are to be gained as well. In addition to being drawn down in dry periods to support consumptive uses such as drinking water, groundwater supplies can be used to support stream flows. By shifting consumptive demands away from surface water and onto stored underground supplies, or even pumping groundwater to augment surface water flows, people can meet consumptive needs while protecting scarce surface water supplies for other purposes, such as aquatic and riparian habitat or aesthetic and recreational uses.

Conjunctive management is generally intended to enhance the protection of water quality also. Surface water quality may suffer during dry periods because of increased wastewater concentrations or the growth of aquatic vegetation, and may be improved by a conjunctive management operation that leaves more water in the stream. Replenishment of groundwater supplies when surface water is abundant can reduce overdraft-related threats to groundwater quality.

Finally, coordinating the use of surface and underground water resources can reduce the size of water infrastructure systems and their financial and environmental costs. Relying on an underlying aquifer for a portion of water storage, or using wells to supplement dry-period surface flows, decreases the needed total capacity of above-ground water storage and distribution facilities. Surface-water facilities can therefore be sized to meet many, but not all, peak or emergency needs. The environmental disruption of additional surface construction and permanent structures is lessened. Substantial financial savings can be realized by avoiding the cost of building and maintaining facilities that would be used only sporadically.

How It's Done: Methods of Conjunctive Management

As suggested in the preceding section, conjunctive management generally is intended to achieve these purposes by capturing surplus surface water supplies and storing them underground in aquifers. The groundwater basin provides a regulatory storage medium that helps to smooth out the greater variability of water demands and surface water supplies. The underground storage constitutes a nonevaporating groundwater "bank" that can be tapped during periods of inadequate stream flow to sustain consumptive uses or supplement stream flows. This bank can even be "overdrawn" during periods of extreme or prolonged surface water shortage, if balance is restored to the system in some future wetter period.

Overall, surface water supplies and storage facilities and underground water supplies and storage capacity are treated as components of a single system, and water needs are met by shifting mixes of surface and groundwater supplies determined by their relative availability. Although the precise means by which that is accomplished will differ from location to location according to physical and other factors, we can summarize the primary methods by which conjunctive management is accomplished.

Storing Groundwater via Direct Recharge

An essential element of any conjunctive management program is how water is moved from the surface to underground storage. This is known as groundwater replenishment or recharge. It may be accomplished directly, by "natural" or "artificial" recharge,[3] or indirectly by a method we discuss later known as "in–lieu" replenishment.

Direct recharge is the name given to deliberate attempts to move water from surface water bodies to underground aquifers. In other words, it is the effort to augment the replenishment that occurs when rainfall or runoff on the surface seeps underground, or percolates.

Since its inception, conjunctive management has been practiced using natural stream or lake beds as the recharge media, provided the topography and soil types were favorable. Areas with flatter terrain and permeable—for example, sand and gravel—stream bottoms could take advantage of a relatively slower rate of movement of stream flows to increase the amount of water seeping underground. Groundwater recharge by this method can be enhanced by using a surface water impoundment, such as a flow-regulating dam, to control the volume and flow of water in the stream channel. In this manner, a greater proportion of the stream flow sinks underground along the streambed.

Another common method of groundwater recharge involves diverting water from stream channels into broad plains or ponds adjacent to the channel, where the water would collect to percolate through the soil (Todd and Priestaf 1997, *140–141*). This practice gave rise to the deliberate construction of "spreading basins" with permeable bottoms of sand and gravel, and in-channel diversion facilities such as sand dikes or, later, low inflatable dams to shunt flows out of a stream channel into the basins.

Figure 2-1 is an aerial-view representation of a conjunctive management operation that employs both recharge methods. The figure shows a surface-water impoundment facility: a dam and reservoir. The dam may have been constructed and operated as a flood-control facility, but not necessarily. Surface water gathering behind the dam is released at a controlled rate to regulate flows downstream. If the stream channel below the dam is permeable, if it is composed of gravel and loose sand, some of the water released through the dam will percolate through the streambed into underground storage. Any water not percolating through the streambed would continue downstream, eventually reaching another stream, the ocean, or other outlet, thus effectively being "lost" to the water users of the area in the figure. The groundwater recharge objective would be to re-regulate flows through the dam to a rate that maximizes the amount of water percolating through the streambed into the local aquifer; in turn this minimizes the amount of streamflow that continues downstream, and is lost for storage opportunities.

It is of course possible that the area represented in Figure 2-1 occasionally receives too much precipitation and runoff for releases through the dam to be held to such a low rate. Larger flows through the dam may be needed. If local water users wish to capture those flows, too, rather than let them continue downstream, they may choose to invest in the construction of percolation ponds or spreading basins adjacent to the stream channel below the dam. During high-flow periods, some portion of the larger releases from the dam is diverted into these offstream ponds. This adds to the amount of groundwater replenishment that can be accomplished through streambed percolation alone.

Simple though it is, Figure 2-1 represents essential elements of a common approach to conjunctive management. Surface water supplies and the storage facilities represented by the dam and reservoir are operated in coordination with underground storage. Temporarily impounding surplus surface flows behind the dam can protect downstream areas from flooding, and allow for controlled releases of the stored water to maximize the amount that is conserved as stored groundwater. During high-flow periods, groundwater recharge can occur through both the streambed and the percolation ponds. In low-flow periods, it may be possible to release water through the dam gradually enough to accomplish recharge through the streambed alone.

Maintaining a high rate of recharge in percolation ponds or spreading basins is critical to achieving the maximum volume of stored water. Because

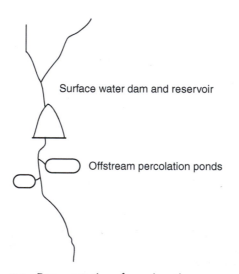

Figure 2-1. Representation of a conjunctive-use operation

the size of a spreading basin is fixed in the short term, the amount of water that is recharged to the groundwater basin is a function of how quickly water moves underground. Fine silts in surface water settle at the bottom of a spreading basin, clogging up the spaces through which the water is supposed to pass. This reduces the rate of percolation, which can create multiple problems. Not only does the amount of water received by the aquifer decline, but in addition the spreading basins do not empty quickly enough. Releases from upstream surface reservoirs may have to be delayed, which can have flood-control implications, and in the meantime, the quality of standing water in the percolation basins can deteriorate. Therefore the spreading basins in some sophisticated contemporary conjunctive management programs are maintained or "cleaned" when not in use. Basin managers periodically scour the basin bottoms to restore their permeability and enhance the recharge rate.

Percolation is not the only means by which surface water can be stored underground. In many locations, the composition of surface or subsurface soils may not be conducive to groundwater replenishment by percolation. Extremely fine, claylike soils, for instance, have poor permeability and impede the passage of water from the surface to an underground aquifer. In other locations, aquifers are composed of rock rather than sand or gravel. Under these circumstances, a conjunctive management operation may involve injection wells—wells drilled from the surface to the aquifer, but equipped with pumps designed to drive water down into the ground rather than lift it to the surface.

Because it entails additional equipment and energy costs, groundwater replenishment via injection is more expensive than percolation. The technique is nevertheless employed in some locations where the percolation approach is infeasible.

Alternative Sources of Replenishment Water

To this point, we have presented groundwater replenishment in terms of capturing and storing natural surface water flows. Aquifers can, however, be recharged using other water sources.

Highly treated wastewater—also known as reclaimed or recycled water—can be and has been used to recharge groundwater, alone or in combination with surface water supplies. Using treated wastewater for groundwater replenishment also represents a recognition of complementarities. Treated wastewater often is disposed of anyway; directing it to basin replenishment provides an avenue for reuse. In addition, where recharge can be accomplished through percolation, it adds another step of filtration to the water treatment process.

Desalinated ocean water could also be stored underground. Ocean desalination remains comparatively expensive, however, and its availability is obviously limited geographically in ways that wastewater is not.

Recovering Stored Groundwater

Groundwater replenishment, by itself, does not constitute conjunctive management. Another crucial aspect is the subsequent extraction and use of water supplies that were stored underground.

Periods of below-normal precipitation, and more extreme periods of drought, reduce the availability of surface water supplies. Scarce surface water flows create two sets of problems: increased difficulties of allocating those diminished quantities among competing consumptive uses, and the consequences for nonconsumptive uses such as aquatic and riparian species and habitat or human aesthetic and recreational enjoyment. Stored groundwater can be drawn on to supply consumptive needs, and to supplement the instream flows needed to sustain ecological, aesthetic, and recreational values.

In most conjunctive management programs, stored groundwater is extracted through existing wells. In some locations, the implementation of conjunctive management has required the installation of additional groundwater pumping capacity in order to make larger-volume shifts from surface water to groundwater possible.

In other cases, instream flows are supplemented through the natural flow of water between aquifers and streambeds. Aquifers that are hydrologically connected to surface streams are recharged at various times and locations so as to maximize return flows to surface streams during periods of peak water demand.

Sustaining sufficient groundwater extractions throughout an extended drought often results in drawing out more water than was stored during the preceding period of recharge—in other words, deliberately overdrafting the aquifer. A conjunctive management program typically entails making up such overdrafts during the next wet period, in addition to storing more water underground to guard against the next dry period or drought. Planned overdrafting thus feeds back into the scale of facilities needed for surface water diversion and impoundment.

Indirect ("In-Lieu") Groundwater Replenishment

In areas where water users have access both to groundwater wells and surface water connections, a simpler method of achieving conjunctive management is feasible. It involves adjusting water users' relative demands upon surface and groundwater sources, and is known as indirect or "in-lieu" groundwater replenishment.

An in-lieu program works in the following manner: during relatively wetter periods, water users are encouraged to meet a greater share of their needs from surface sources in relatively wetter periods and refrain from pumping groundwater. In drier periods, they are encouraged to shift to greater reliance on groundwater pumping.

The term "in-lieu replenishment" was coined to capture the fact that increased surface water consumption and decreased groundwater extraction have a combined effect on the aquifer that is similar to direct recharge. By foregoing pumping, and taking surface water in lieu of groundwater, groundwater users leave water underground that they otherwise would have withdrawn. It accumulates and carries over to future dry periods when it is needed. The net effect to the aquifer is the same as if water users had withdrawn groundwater while basin managers used the surplus surface water for direct recharge.

There are some operational advantages to in-lieu replenishment, however, as it is accomplished without devoting as much physical effort to the capture, controlled release, and percolation of surplus surface water into the ground. For this reason, a water agency pursuing a conjunctive management program in such a setting may even offer surplus surface water at a reduced or subsidized rate to encourage water users to employ it instead of their wells.[4]

Recent Trends Contributing to Conjunctive Management

As discussed in Chapter 1, calls for greater implementation of conjunctive management have increased steadily from the 1950s through the 1990s, especially in the American West. Besides the mounting pressure from new and increasing water demands—population growth, urbanization, agriculture intensification, and environmental protection—financial considerations and increased concern to mitigate groundwater overdraft are also driving the rising attention to conjunctive management.

Constraints on Construction and Operation of Surface Water Storage Facilities

Through much of the twentieth century in the American Southwest, demands for increased water storage and distribution capacity were met by building vast surface water storage and conveyance systems. For several reasons, the prospects for meeting additional water demands via additional construction have dimmed, to say the least (Reisner 1994). By the outset of the twenty-first century, at least as much attention was given to the prospect of dismantling dams as to building new ones. The Arizona legislature, for instance, adopted a resolution in February 2001 requesting that the United States Congress oppose any efforts to breach or remove Glen Canyon Dam or to drain Lake Powell.

The same urbanization and agricultural development processes that boosted water demands also raised land prices, so it is now far more expensive to acquire land for dams and reservoirs and rights of way for aqueducts and pipelines. In some areas, estimates of the cost per acre-foot of additional surface water storage have approached the cost of desalting ocean water—long regarded as the benchmark against which more conventional water development efforts have been measured. For example, according to estimates by the Contra Costa Water District, enlarging the Los Vaqueros reservoir in the East Bay area of California from its current 100,000 acre-feet capacity to 500,000 acre-feet could cost $900 million (Taugher 2002). The financial impacts associated with the construction of additional surface storage to meet occasional dry-year peak demands are especially high.[5] Despite the costs, suppliers still consider the option of constructing additional surface water storage and distribution given the tremendous political pressure to ensure water supply reliability.

In the meantime, the financial situation of the federal government from the late 1970s through the mid-1990s, and of state governments through the 1980s, reduced federal and state officials' ability and willingness to fund new facilities on the scale needed to meet new demands using the old methods. The same financial constraints limited funding for ongoing maintenance of surface water facilities, some of which are beginning to weaken or to lose capacity as a result of silt accumulation.

Had the financial feasibility of building additional surface water storage and conveyance projects not changed, environmental regulations adopted since the early 1970s would nevertheless have slowed or stopped many dam or pipeline projects at the planning stage. The regulatory restraints on any activity that will impound instream flows and affect aesthetic or ecological values have accumulated (Reisner and Bates 1990). The few remaining streams or canyons where a dam might be sited for water supply development are also likely to be protected as habitats for threatened or endangered species or for other purposes.

As construction of additional surface water storage has faded, existing facilities are being reoriented to serve new and multiple purposes. The importance of the flood-control functions of dams and reservoirs rises with the concentration of people and businesses. Some dams built with agricultural water conservation purposes in mind originally are now upstream from fast-growing cities, with the protection of hundreds of thousands of lives and billions of dollars in property at stake in their operation. In other locations, dams built primarily for flood-control purposes are being reoriented to serve water conservation and water supply needs. These are delicate and difficult balances to strike. As noted earlier, operating a surface water project for flood control purposes requires maintaining empty reservoir space in the same wet season during which operating the project for water supply requires filling it up. Rapid releases of water to preserve adequate flood-control space are generally incompatible with the effort to conserve water supplies and make them last through the dry season.

Over the last quarter-century or so in the American Southwest, the value of surface water supplies for purposes other than irrigation and drinking has risen (Colby 1989). Operators of dams and reservoirs are under greater legal and political pressure to operate them for species preservation and habitat restoration purposes. Two landmark pieces of legislation were passed in 1992, reflecting this trend. The Central Valley Project Improvement Act, mentioned in Chapter 1, required the U.S. Bureau of Reclamation to operate the irrigation-oriented federal Central Valley Project in ways that now also ensure water supplies for species and habitat protection in the San Francisco Bay–San Joaquin Delta region. The California Department of Water Resources has begun reoperating the State Water Project for the same reasons (California Department of Water Resources 1995).

The Grand Canyon Protection Act required the Department of Interior to operate the Glen Canyon Dam to protect environmental, cultural, and recreational resources in Grand Canyon National Park and Glen Canyon National Recreation Area and to establish long-term operating procedures for the dam that would protect such resources (P.L. 105–575).[6] Since the act's passage, the Bureau accounts explicitly for environmental considerations in its annual operating plan for Colorado River system reservoirs. Furthermore, as a consequence of the U.S. Fish and Wildlife Service's listing of critical habitats for four endangered species along the Lower Colorado River in 1994, the Bureau spearheaded the Lower Colorado River Multi-Species Conservation Program (LCRMSCP). The program—developed through a consortium of federal agencies; Arizona, California, and Nevada state and local agencies; Native American tribes, water and hydroelectric agencies, and numerous nonprofit organizations—has as its stated purposes the conservation of habitat and the recovery of threatened and endangered species while also accommodating current water diversions and power production (LCRMSCP 2001). The program is likely to have further effects on the Bureau's management of the lower Colorado River.

These new operating requirements of surface water facilities may be contradictory to the requirements of water supply and storage for human consumptive uses (Krautkraemer 1992, *152*). Water released from reservoirs in the spring to support streamflow for spawning fish and nesting birds, for example, is not available to be released in the late summer for thirsty people or withering crops.

Throughout the West, rising consumptive and nonconsumptive water demands are coinciding with rising opposition to the construction of additional surface water facilities or the enlargement of existing ones. Existing facilities are vulnerable to silt accumulation and other manifestations of neglected maintenance, or are being recommitted to multiple purposes, or both. Increased emphasis on conjunctive management, which shifts the primary emphasis of dry-period storage from surface facilities to underground basins, is clearly a result of these trends. The connection among these trends was drawn clearly in the most recent update of the California Water Plan, where the state's Department of Water Resources stated:

Efficient use of surface and ground water through conjunctive use programs has become an extremely important water management tool. Such programs are generally less costly and cause fewer adverse environmental impacts than traditional surface water projects because they increase the efficiency of existing supply systems without requiring major facility additions. (California Department of Water Resources 1994a, *12*)

Financial Advantages of Underground Water Storage

In this context, recognition of the economic value of groundwater basins as reservoirs for water storage has grown. Storing water underground has become a financially, as well as environmentally, attractive alternative to new dams and reservoirs.

The value of a groundwater basin can even be measured in terms of the avoided cost of an equivalent quantity of surface storage capacity (Blomquist 1992), and those avoided costs are greatest with respect to storage capacity to meet occasional peak or emergency demands. Taking advantage of this "buffer value" of groundwater storage is a way of recognizing what may be its highest economic use (National Research Council 1997, *60*).

In a 1995 review of options for implementing the provisions of the Central Valley Project Improvement Act, the U.S. Bureau of Reclamation identified increased conjunctive management as the least-cost alternative. Estimates of the cost of water yielded through conjunctive-use operations were in the range of $60 to $120 per acre-foot, compared with a range of $300 to nearly $3,000 per acre-foot from the construction of additional surface storage facilities.

The 1995 Integrated Resources Plan for the Metropolitan Water District of Southern California identified conjunctive management, at approximately $300 per acre-foot, as the second least expensive option for obtaining additional water supplies (Metropolitan Water District of Southern California 1996a). The only less expensive option was moving additional Colorado River water through MWD's aqueduct—an option that has since become physically infeasible, since by the end of the 1990s the district was operating the aqueduct at full capacity.

A similar finding resulted from a study conducted at approximately the same time for a large urban water district—the Bay Area's East Bay Municipal Utility District, or EBMUD. EBMUD was contemplating additional surface water storage capacity to guard against drought and earthquake emergencies—in other words, capacity to store water that would be needed only occasionally and unpredictably. Based on a computer model of EBMUD's current facilities and future plans, Fisher et al. (1995, *304*) found "that a combination of conjunctive use and water marketing is well over an order of magnitude cheaper than the traditional alternative of constructing new storage capacity." They estimated the cost of the additional surface storage options being considered by EBMUD at $1,000–$12,000 per acre-foot. Their estimates of the costs of three options based on conjunctive management ranged between $5 per acre-foot—

for a water-banking option using existing wells and an EBMUD subsidy to the well owners to extract more groundwater in dry years so EBMUD could draw surface reservoirs down farther—and $175 per acre-foot—for a direct-purchase option in which EBMUD would pay for pumped groundwater as a supplemental supply when needed. Adding in the cost of some pricing measures to encourage further residential water conservation, enhanced water supply reliability via a combination of demand reduction and conjunctive management cost an estimated $50–$350 per acre-foot (Fisher et al. 1995).

Making Lemons into Lemonade: Reducing Overdraft and Facilitating Water Reuse

In the latter half of the twentieth century, groundwater use accelerated in the western United States following the introduction of turbine pumps. Groundwater supplies, of course, did not. A result has been substantial groundwater overdraft in many locations in the West. In Arizona, annual overdraft in the Phoenix and Tucson metropolitan areas is an estimated 350,000 acre-feet per year (Arizona Department of Water Resources 2001b). Groundwater overdraft per year is an estimated 1.3 million acre-feet in California alone (California Department of Water Resources 1994a, *12*). More than 15 million acre-feet of groundwater were removed from storage in the San Joaquin Valley during the 1987–1992 drought, with accompanying water level declines of 20–100 feet and some incidents of subsidence (Hauge 1994, *12–13*).

The overdraft problem in the West has been the subject of numerous studies and attempted policy responses. Whatever other concerns may have motivated the development and adoption of Arizona's 1980 Groundwater Management Act, discussed further in Chapter 5, it is clear that at least one of them was the accumulating groundwater overdraft within the state, which threatened to place most groundwater beyond the economically feasible reach of the farms, mines, and cities that depended on it. Arizona now devotes state funds to acquiring and storing unused Colorado River water underground, which not only protects the state's water allocation from the river but is also filling dewatered underground storage space. During the 1980s, U.S. Representative George Miller (D–CA) repeatedly introduced the Reclamation States Groundwater Recovery and Management Act, which would withhold federal water project funds from any western state that did not adopt stricter controls over groundwater withdrawals to arrest overdrafting (U.S. ACIR 1991). The accumulated overdraft in portions of the Ogallala aquifer, which underlies parts of seven western states, has been a subject of study and concern for more than a quarter of a century.

The extent of accumulated groundwater overdraft in the West is a significant resource management problem. It is also an opportunity. This rapidly developing region, which finds it nearly impossible to build additional above-ground water storage facilities, also rests atop a cavernous amount of available underground storage capacity. According to a study commissioned by an association of groundwater management agencies in California, more than 21.5 million acre-feet of groundwater storage capacity is currently available in the southern half of the state alone (AGWA 2000, 2). Not all of that storage capacity is located near the areas of greatest water demand, but as Arizona has demonstrated, it is possible to implement conjunctive management on an interbasin scale. Water users in one location could store water underground in excess of their own needs, and sell the surplus to water users in another location (Bachman et al. 1997, 65).

If ever need met opportunity, it has happened in the case of conjunctive management in the West. Occasional surpluses in surface water supplies must be captured and stored somewhere if dry-period consumptive and nonconsumptive needs are to be met. Dewatered aquifers need to be replenished if further damage from compaction, subsidence, and water quality degradation are to be avoided. Replenishment of those aquifers also raises underground water levels, which reduces pumping lifts and the associated energy costs (Association of Ground Water Agencies 2000, 4).

Among the water sources that could refill those overdrafted aquifers is treated wastewater. Here too, need meets opportunity. Growing populations produce more wastewater, which requires treatment to meet state and federal discharge standards. As of 1993, 380,000 acre-feet per year of treated wastewater were being reused in California alone, and a survey of water agencies projected that amount would rise to 1.3 million acre-feet by the year 2010 (Mills 1994, 32). In Arizona, as of 1994, 120,000 acre-feet per year of reclaimed water were being reused, primarily to irrigate large turf facilities such as golf courses and cemeteries (Arizona Department of Water Resources 2001b).

Although there are plenty of nonpotable uses for recycled water—landscape and agricultural irrigation, and some industrial processes—those uses may not grow quickly enough to accommodate the projected tripling of treated water production. Use of purified water to meet at least some potable demand would certainly direct that water to its highest-valued use. Yet, consumers are clearly reluctant to integrate recycled wastewater directly into their drinking water systems. The indirect route into the potable water system is through groundwater recharge. As of 1993, more than half of the recycled water in California was being used for groundwater replenishment (Mills 1994, 32), and groundwater recharge could account for as much as 40 percent of the 1.3 million acre-feet projected for the state in 2010. At least some of the push for increased conjunctive management may therefore be coming from the supply side, as the water reuse industry seeks more valuable end-uses for its growing output. Conjunctive management is an idea that has existed for more than a century, and one whose time has plainly come. The trends in western water supplies, demands, and resource management converge on it with seemingly overwhelming force. It appears to be an economical, environmentally compatible approach to trying to make limited supplies stretch across time and space to meet an expanding array of growing needs.

If the adoption of a policy reform were based solely on its logical fit with the problem it is intended to solve, conjunctive management would be the most widely practiced approach to water resource management in the western United States today. As our research and that of others has found, however, conjunctive management programs dot the western landscape rather than blanketing it. Other factors affect where, when, and how conjunctive management is put into place. We turn now to those.

3

Opportunities and Obstacles for Conjunctive Management

One cannot say that the *idea* of conjunctive management has failed to catch on. For reasons described in Chapter 2, for more than 50 years water resource management experts, including hydrologists, economists, and policy analysts have recognized that conjunctive management can be an economically and environmentally sensitive technique. As mentioned in Chapter 1, however, institutions play a key role in shaping when and where a new water management technique will catch on. In this chapter we explore in more detail the factors that promote, shape, and limit conjunctive management, to complete the foundation for our study of conjunctive management in California, Arizona, and Colorado.

The most important categories are physical and institutional. Physical factors matter because without the appropriate blend of hydrology, geology, surplus water, and distribution systems, conjunctive management is infeasible no matter how supportive the institutional setting. Conversely, no matter how ideal the physical setting, conjunctive management will not occur without a minimal set of supportive institutional arrangements, such as water rights that encourage and protect conjunctive management investments. Also, water organizations, from owners and managers of large surface storage and distributions systems to small municipal water retailers, play a critical role in implementing institutions, and in amassing and coordinating the resources and activities that constitute conjunctive management. Their actions can either hinder or ease the practice of conjunctive management, depending on the physical and institutional setting.

The Physical Setting

A recent issue paper published by the Association of Ground Water Agencies characterized the hydrogeology of groundwater basins as "the single greatest factor influencing" the viability of conjunctive water management in any given location (Association of Ground Water Agencies 1998, 4).[1] According to California Department of Water Resources chief hydrologist Carl J. Hauge (1992, 25), "The best conjunctive use areas have the following properties:

- high amounts of sand and gravel
- high percolation rates
- high well yields
- high storage capacity
- no continuous clay layers."

As Hauge suggests, the composition of materials in an aquifer can affect the volume of water that can be recharged and stored. In addition, the soil layers between the land surface and water table, known as the vadose zone, can contain sediments that keep water from penetrating all the way to the aquifer. These factors also influence the ability to recover stored water quickly and cost-effectively. If an aquifer has poor percolation rates, low storage capacity, or low well yields, conjunctive management may not be feasible for that aquifer even if sufficient surface water supplies are available (Hauge 1992). Hydrogeologic properties of aquifers can vary across sites in a basin and across sub-basins.

The groundwater basins underlying Arizona's two main population centers, Phoenix and Tucson, demonstrate the importance of hydrogeologic factors in facilitating conjunctive water management. Although both areas feature gently sloping alluvial plains with relatively high amounts of sand and gravel soils, direct recharge and recovery programs in the Phoenix area typically are larger and are more widespread than in Tucson because of the faster percolation rates and higher well yields of the Phoenix basins.

Surface water–groundwater interaction also affects the practical application of conjunctive use. Insufficient interaction between the surface and the aquifer—for instance, where a layer of clay seals off an aquifer from the land surface—will make introduction of water into the aquifer difficult and require the use of more expensive methods such as injection wells. At the other extreme, an aquifer may be too closely related to the surface. In portions of the Sacramento River Valley, water moves so readily from the surface to underground and back again that it is hard to store water underground long enough to be beneficial for purposes of the management program (Natural Heritage Institute 1998). In other words, water stored underground does not "stay put," emerging too soon as rising water in valley streambeds.[2]

Surface water supplies and adequate groundwater storage sites must physically coexist for conjunctive management to be implemented. Hauge's points concerning the relative locations of water demands, surface water supplies, and underground storage capacity are especially important in the American Southwest. Numerous sites in the Southwest are well suited to conjunctive management, but water demand, surface supplies, underground capacity, and a desirable amount of surface–ground interaction are not automatically found together.[3]

Where these elements do not naturally coexist, they must be brought together to make conjunctive management work. Thus, in addition to appropriate hydrogeology, necessary components of a conjunctive management program include surface water supply distribution facilities, space for recharge facilities, sufficient underground storage capacity, wells for retrieval of stored water, and infrastructure to deliver stored water supplies.

It would be difficult to overstate the importance of surface water availability, groundwater basin characteristics, and their physical proximity in facilitating conjunctive management. Conjunctive management has long been viewed as a promising tool for meeting the conflicting agricultural, urban, and environmental demands of California's Central Valley, for instance. In the northern portion of the valley, surplus surface water supplies are readily available in the Sacramento River region for conjunctive management, but underground storage sites are limited (Natural Heritage Institute 1998). In the southern portion of the valley, the San Joaquin River region possesses enormous underground storage potential but often lacks surplus surface water supplies.[4]

As we have already observed, few surplus surface water supplies exist in the western United States. Likewise, many suitable groundwater basins are located far from areas that have access to infrastructure for water transportation, storage, and recovery. Population centers in Arizona, California, Colorado, Nevada, New Mexico, and Utah are clustered in a handful of locations. Most surface supplies of any magnitude are already tapped to meet the daily requirements of those developed areas. Furthermore, some of the largest and most suitable underground basins in those states are so far from the developed areas that transporting water to and from underground storage adds dramatically to the cost. Physical disparities such as these limit the opportunities for engaging in conjunctive water management.

Institutional Factors

The practice of conjunctive management requires a great deal of joint and coordinated effort among individuals and organizations. For instance, consider the following:

- Conjunctive management projects typically entail the coordination of surface water project facilities such as dams and reservoirs; large-scale distribution systems such as pipelines and canals; finer-scale distribution systems that move surface water to end users such as homes, farms, and other businesses; groundwater supplies and aquifer storage capacity; and extraction wells and pipes that convey groundwater to users.
- Each of these major facility components may be owned, operated, regulated, or otherwise affected by separate public or private organizations, each of

which is governed by rules in the form of laws, regulations, and charter provisions specifying what it may, may not, and must do. In the United States, those governing rules are set at local, state, and national levels.

- The geography and geology of the location will create diverse opportunities and problems—subsidence of overlying lands if groundwater levels are drawn too low, saturation of overlying lands if they are raised too high, migration of contaminants within the aquifer, intrusion of saltwater in coastal aquifers, and so on. The authority and obligation to respond to these issues may reside in the same organizations that operate the water supply facilities, but at least as often they will reside in other, typically public and often local, organizations.
- The surface ecology and economy of that location will present opportunities and limits on the management of surface water supplies—stream flows to be maintained, flood control needs to be met, water quality standards to be observed, consumptive uses to be satisfied, and the financial resources or cost constraints to be taken into account while doing so. These issues stretch across the agendas of other public and private organizations, with the public organizations reaching to the national level to fulfill some responsibilities.
- Finally, extensive monitoring is required—of weather conditions, water conditions above and below ground, consumptive use requirements, species and habitat conditions, and so forth—to implement and adapt the conjunctive management project successfully.

At every step in the development and implementation of a conjunctive management project are found institutional issues to confront, respond to, take advantage of, or try to minimize.

If rules and organizational arrangements do not facilitate the coordination of actions necessary to divert, impound, recharge, store, protect, and extract water for conjunctive management, that coordination is less likely to occur. In Chapter 1 we alluded to this relationship between institutions and the prospects for conjunctive management, and here we take water rights as one example of institutional arrangements affecting water users' actions and discuss in some detail the effects of those rights on the prospects for conjunctive management and the forms that it may take. We anticipate that, all other things being equal, more completely specified water rights will be conducive to resolving water management problems. Incomplete, or unclear and therefore contested, rights will exacerbate the management problems water users face. With regard to conjunctive management in particular, dual or multiple water-rights systems—such as one set of rules for surface water and another for groundwater—raise substantially the costs of reaching agreements and implementing projects.

Property rights authorize actions with respect to a thing, and bar actions by others or impose duties on them (Commons 1968). The authorizations and duties established by water-rights laws constitute the basic institutional infrastructure by which a water resource is governed. In most western states, for instance, water-rights laws authorize the owners of land to access and withdraw underlying groundwater, subject to specific regulations and guidelines, and impose duties on those landowners not to interfere with one another's withdrawals.

The property rights an individual enjoys in relation to a resource can vary. A complete bundle of rights consists of rights of access, withdrawal, management, exclusion, and transfer (Schlager and Ostrom 1992). This complete bundle of rights is associated with private property. The rights of access and withdrawal give an individual the authority to access property and to make use of it. The right of management gives an individual the authority to make choices over how the property will be used. The rights of exclusion and transfer allow an individual to control access to property and to sell, lease, or bequeath it. These rights are cumulative, that is, the right of withdrawal implies the right of access, and the right of management implies the rights of access and withdrawal, and so forth (Schlager and Ostrom 1992). The rights and duties that an individual exercises in relation to a resource depend on the bundle of rights that the individual holds.

Western states generally have defined well-specified rights in native surface waters using the prior appropriation doctrine. The prior appropriation doctrine allocates water on the basis of seniority, or "first in time, first in right." Usually through a process of adjudication, individuals are granted rights to use up to a specific amount of water. When water shortages occur, the rights of those who have held rights for the longest time are completely satisfied, while those who have held rights for a lesser duration may not receive any water. Under the prior appropriation doctrine, individuals are granted fairly well defined and transferable rights, making it feasible to defend, modify, and even sell their surface water rights.[5]

On the other hand, whereas some western states apply the prior appropriation doctrine to groundwater, other states follow variations of the "beneficial use" doctrine, allowing overlying landowners to pump unspecified amounts of groundwater as long as they do not engage in wasteful uses or interfere with the rights of other overlying owners (Z. Smith 1989). Because this doctrine does not authorize individuals to control specific amounts of water, groundwater is more nearly an open-access resource for overlying landowners.

An individual's bundle of rights with respect to the use of surface water may, and in many instances does, differ from his or her bundle of rights with respect to groundwater, even though the underground and surface water systems are hydrologically interconnected. To complicate the setting even further, most individuals and organizations rely on a third source of water—project water transported from some other location and delivered by a public or private project operator. Instead of falling within the categories of prior appropriation or beneficial use, project water is usually governed by a contract agreed on between the project operator and project water recipients. The terms of such contracts vary considerably. For instance, the Northern Colorado Water Conservancy District (NCWCD), which owns and operates the Colorado–Big Thompson project, allows members to buy, lease, and sell their allotments of project water to one another. The NCWCD also does not retain ownership of the water after the members use it; water that is discharged from members' properties and flows into the South Platte River becomes part of the river system. In the Arkansas River watershed, a different project operator—the Southeastern Colorado Water Conservancy District—does not allow its members to buy, sell, lease, or otherwise transfer their project water, even to other members, and the District retains ownership of all discharges or return flows, which it again leases out to members. In the practice of conjunctive management, then, the different water sources on which users rely are usually governed by different systems of rights, and this complicates the implementation of conjunctive management considerably.

Differing property rights bundles create different incentives for individuals. Garrett Hardin (1968) popularized this critical insight through the model of the tragedy of the commons. Individuals jointly exercising the minimal rights of access and withdrawal find themselves in a race to ruin. Capture and consumption are rewarded, restraint and conservation penalized: individuals gain the full benefit of what they harvest, whereas those who conserve find the benefits created by their actions enjoyed by others. Hardin argued that to avoid disaster, property rights systems must be changed.

Property rights that allow individuals to experience both the benefits and costs of their actions are likely to promote behavior substantially different from that resulting from systems such as Hardin's commons that allow individuals to capture benefits and shed costs. Changes in water rights have been clearly associated with changes in behavior among western water users. Before groundwater in Colorado was brought under that state's system of appropriative rights, discussed further in Chapter 6, groundwater pumpers built wells and pumped with few restrictions. Pumpers gained the full benefit of their actions while imposing the costs on others, particularly surface water rights holders as stream flows declined as a result of groundwater pumping. Shifting groundwater rights to the appropriation system required Colorado pumpers to bear the costs of their actions. Although pumpers were granted quantified rights to a portion of groundwater, they could pump only if their rights were in priority; otherwise, they were liable to senior rights holders for harm created by their pumping. Before the property rights change, the number of wells in eastern Colorado grew geometrically; after the change, the number of wells stabilized and eventually began to decline.

Property rights also influence individuals' expectations about the future. The expectations of those who possess rights of exclusion will vary considerably from those of the individuals who do not. Without the right of exclusion, individuals know that others may capture the future benefits of investments they make in their property, and so they are less likely to make those investments. With an enforceable right of exclusion, individuals are more likely to undertake long-term investments. If individuals also hold rights of transfer, their incentive to engage in long-term investments is even greater, as they can capitalize on investments either by enjoying the future benefits themselves or by transferring the property to someone else for a price that reflects those benefits.

By influencing incentives and expectations, property rights affect the use and practice of conjunctive management in some obvious and some not so obvious ways. At a minimum, a certain level of control over surface, project, and groundwater supplies is necessary for conjunctive management to occur. In direct recharge projects, conjunctive management participants need assurances that the water they store underground for their own benefit will not be taken by others. When groundwater is governed by the beneficial use doctrine, those assurances do not exist and developing them can be costly. All else being equal, the greater the number of water users who must negotiate water storage and retrieval assurances, or coordinate the timing and quantity of water going into conjunctive management projects, the greater the costs of implementing conjunctive management.

In in-lieu recharge programs, water users forego pumping groundwater they would normally use and rely on surface water instead when it is comparatively plentiful. These water users expect to rely more heavily on the aquifer in years when surplus surface water is not available, recapturing the groundwater they had earlier left alone. Where groundwater is treated as an open-access resource or groundwater rights are not quantified, users lack the guarantee that others in the basin will not consume the groundwater supplies that the participants in the in-lieu program expect to have available in the future. Under these settings, in-lieu program operators must negotiate or coordinate with all pumpers in the basin to devise assurances that they will have access in dry years to the groundwater they did not use.

When the dry years occur, participants in an in-lieu program will pump more intensively to make up for reduced surface water deliveries. Others overlying the same basin may want assurances that the in-lieu participants' increased pumping will not harm them. Providing that assurance will also require negotiation and bargaining. We would expect to find that the quantification of water-use rights would facilitate the development of the assurances that water users need to cooperate in conjunctive management.

Water rights laws that extend quantified rights to the storage and recovery of water also help provide the assurances needed by prospective participants in a conjunctive management project (U.S. Advisory Commission on Intergovernmental Relations 1991). Rights to manage stored water, exclude others from capturing it, and even transfer stored water to others help assure participants that they will maintain control of water they commit to a conjunctive management project, and thus be able to recover benefits from the project.[6]

Overall, then, the more complete the bundle of rights that individuals hold in different types of water the more likely they are to participate in conjunctive management. When water rights are incomplete, users are less likely to exercise the restraint involved in storing water as part of a conjunctive management program. When water rights are unclear or when differing and contestable claims arise in relation to the same water, users bear the costs of resolving conflicts and negotiating and/or enforcing solutions about who may do what in relation to which aspects of the resource. And when users hold differing bundles of rights (clear or unclear, complete or incomplete) with respect to various aspects of the resource—surface versus groundwater, native versus imported project water, water flows versus water in storage—the costs of coordinating conjunctive management projects can be daunting.

At least one caveat to this argument about clarity and completeness, however, is appropriate here. One of the signal advantages of conjunctive management is flexibility. The most productive conjunctive management projects are those that allow water managers to switch relatively easily between surface water and groundwater to maximize availability and/or minimize cost; in other words, that allow project managers to treat surface water and groundwater as essentially interchangeable. Although more complete bundles of property rights provide the security that encourages individuals to undertake beneficial investments, security of property rights may undermine the flexibility of conjunctive management.

Conjunctive management programs often treat the aquifer as a unit—a collective good—valued for its overall ability to store and yield water. The yield and storage capacity of the aquifer may be managed for the benefit of overlying landowners or of permit holders, but also perhaps for the benefit of other, nonoverlying water users and even other species. Where institutional arrangements assign individual property rights to the groundwater yield or storage capacity of an aquifer, but that aquifer is managed using a conjunctive management program, the potential for conflict exists. The ability of individual rights holders to use judicial or administrative proceedings to assert their rights to a guaranteed volume or rate of groundwater yield can constrain or thwart efforts to replenish a basin and draw down the basin during dry periods.

This sort of conflict may not make conjunctive management impossible, but the property rights individuals hold can be expected to affect the institutional and organizational forms that are created to undertake conjunctive management. For instance, the overlying landowners of a basin governed by the beneficial use doctrine may create a governing organization and transfer to it their rights of management of groundwater. Rather than allowing each individual landowner to be able to pump as much as he or she can put to beneficial use, the transfer of such rights to an overlying jurisdiction would allow it to coordinate the landowners' groundwater use. With such a change in property rights, conjunctive management becomes feasible. By acquiring the right of management, the overlying jurisdiction possesses significant flexibility in pursuing conjunctive management. Presumably, as long as it makes enough water available to landowners, the jurisdiction may actively use the basin for the long-term storage of surplus water, which can be drawn on in times of drought.

Assume instead that the landowners choose to switch from the beneficial use doctrine for governing their basin to defining and allocating quantified rights in shares of groundwater. As we suggested in the preceding, such a change in rights would make conjunctive management a more attractive option than it was under the beneficial use doctrine. Groundwater rights holders would be assured that water they stored in the basin would be available to them when they wanted to draw on it, because their fellow rights holders could withdraw no more than the amount of water to which they held rights. However, the active use of the basin as a long-term storage unit may be more limited, as increased withdrawals during dry periods may substantially drop the water table and adversely affect water users' abilities to access and pump their water. The conjunctive management under this scenario would differ from the one discussed earlier. The use of the basin for long-term storage may be more limited because of the negative effects on the rights of individual water users of raising and lowering the water table.

Finally, consider a scenario in which groundwater that is hydrologically connected to surface water is brought within the surface water rights system, for example, the prior appropriation doctrine. The quantification of pumping rights and their inclusion in the same seniority system as surface water rights helps to make rights to both water sources secure enough to encourage investment in conjunctive management. Conjunctive management under such a scenario, however, is likely to have a different purpose and take a different form than the scenarios described previously in which groundwater is used for storage of surplus surface water. Where surface water and groundwater rights are integrated into one seniority system and surface water rights holders are generally senior to groundwater pumpers, conjunctive management is likely to entail adjusting pumping amounts and groundwater levels to maintain sufficient stream flows. In this scenario, conjunctive management is used to protect the rights of senior surface water rights holders rather than for long-term storage of water to mitigate drought.

The emergence and development of conjunctive management among western states depends not only on whether states have defined sufficiently secure rights in different types of water to encourage individuals to invest in conjunctive management. It also depends on the tradeoffs states have made between security of water rights and flexibility in using different types of water. As different types of water rights affect the security and flexibility to support conjunctive management, they also translate into different forms of conjunctive management. Thus, rights affect not only the likelihood that conjunctive management occurs, but also its structure and operation where it does occur.

Water Management Organizations

Water rights or other institutional rules do not operate in a vacuum; they are typically devised and administered by organizations, and water management choices and activities emerge through the interaction of organizations. In this section we discuss public and private organizations that are likely to be involved in conjunctive management projects—those that capture, convey, manage, store, or sell water—and introduce our expectations about how the practices of and relationships between organizations will likely support, impede, and shape the practice and performance of conjunctive management.

In light of the physical and institutional factors involved in conjunctive management, we can assume that conjunctive management projects will usually require the coordination of multiple organizations. Even if a single organization were responsible for implementing a conjunctive management project, it would likely have to arrange to purchase surplus water from a surface water project, and/or arrange to use the distribution facilities of a surface water project to deliver water to the project site, and acquire or perhaps lease land from another organization for the site of the project. The organization may also choose to lease storage in the project to other organizations, or at a minimum to sell the stored water to other clients.

The involvement of multiple organizations with differing interests and responsibilities raises coordination issues. The interests of groundwater basin managers and surplus surface water suppliers, for example, are likely to diverge. The interests of the purveyor of surplus surface water tend to emphasize (1) minimizing the costs of storing water, (2) minimizing any liability for problems caused by the conjunctive management project, and (3) maximizing the assurance that the stored water can be recovered during dry periods. The interests of a groundwater manager involved in conjunctive management tend to focus on (1) protecting basin water producers and water conditions by securing the best terms and conditions from the surface water supplier, and (2) keeping the terms and conditions regarding recovery of stored water as loose as possible to ensure the availability of groundwater for local producers.

Coordination would be especially vital for large conjunctive management projects in which multiple jurisdictions and functionally specialized organizations are employed so as to take advantage of differing economies of scale (cf. Oakerson 1999). At the same time, coordination entails negotiation and bargaining costs. The benefits of coordinating across jurisdictions of different sizes and functions must be weighed against the costs of coordinating across these jurisdictions to produce the conjunctive management program.

All interests in a conjunctive management program will have bargaining limits. Presumably, a rational surface water supplier could not be driven to pay costs in excess of the next best alternative, either storage in another basin or the cost of surface storage. A rational basin manager would not agree to a storage-

and-recovery program that increases the costs or reduces the availability of local groundwater supplies during a dry period. A considerable negotiating range nevertheless should exist within those limits. Because either party is answerable to other constituents or directors, the time and effort expended finding an agreement that both will accept and implement can be considerable.

Coordination problems often center on devising equitable allocations of benefits and costs among participants involved in conjunctive management. Nowhere are coordination problems more likely to emerge than around financial issues. Although conjunctive management is often less expensive than increasing surface water storage capacity, financial considerations involved in starting up or switching over to a conjunctive management program remain. Recharge site development—which may include the construction of off-channel spreading basins, surface stream channels, or injection and extraction wells—represents one of the primary costs of engaging in conjunctive management. Even if distribution facilities exist relatively close to potential recharge sites, it is unlikely that a given recharge site will have the full complement of pipelines, turnouts, and pumps required to bring water into a site. Conjunctive management projects often require additional procedures and infrastructure to protect water quality.

The benefits of a conjunctive management program may exceed the costs, but water providers may find it difficult to apportion costs equitably among themselves, and some providers may not have the financial capacity to develop the infrastructure needed for a conjunctive water management program. Such coordination issues may be made more difficult by state laws that hinder coordination among jurisdictions. For instance, financing may require interjurisdictional coordination, particularly among small water providers, or the creation of a new jurisdiction to ensure that the area benefiting from the financed infrastructure matches the geographic area responsible for repayment. The opportunities for joint financing or creating new jurisdictions to finance projects may not be permitted by state laws. Water providers in the Tucson area have noted that the opportunities are limited for creating new jurisdictions for financing surface water delivery infrastructure (Tucson AMA Safe-Yield Task Force 2000). Arizona law does not authorize the creation of a special district or jurisdiction specifically for water infrastructure that would span existing jurisdictions.

Just as the relationship between property rights and conjunctive management is two-edged, so too is the relationship between organizations and conjunctive management. Although organizations may be a source of coordination problems making conjunctive management difficult, they could also reduce coordination problems. In fact, for decades states have authorized the formation of special-purpose governments, in the form of water districts, or special authorities, to manage water resource issues that cross multiple jurisdictions. Unlike private water providers, such public entities have taxing authority to support the financing of developing new water management programs. Public and private organizations alike, such as special districts or water users' associations, whose members may encompass multiple water providers, similarly can engage in collective purchasing of water supplies, and can facilitate planning, building, maintaining, and monitoring different aspects of conjunctive management programs. Thus, certain organizational arrangements would likely help lower the costs associated with conjunctive management.

Not only may organizations ease coordination problems, but they may also be designed and used to ease the tradeoffs between the security of property rights and the flexibility of conjunctive management. This could be facilitated, for instance, by a publicly created organization that manages a groundwater basin and acts as a water wholesaler. The organization may actively manage the basin to maximize water supplies while also using water to address the impacts on the property rights of others that occur within its boundaries as a result of conjunctive management. For instance, the Orange County Water District, which manages the basin underlying much of Orange County, California, levies fees on groundwater pumping, which raise the cost and may encourage pumpers to substitute imported surface water for a portion of their needs. The district uses the fee revenue to replenish the groundwater basin, thus supporting groundwater availability during water-scarce years (Blomquist 1992).

Organizations may therefore both hinder and support the practice of conjunctive management. Organizations with overlapping and conflicting interests may or may not overcome their differences to move forward with conjunctive management. On the other hand, some organizations can ease or overcome coordination problems and therefore enhance the opportunities for conjunctive management. Specialized water organizations, whether formed as public organizations through state laws, such as special water management districts, or as private organizations could in theory help facilitate the development of conjunctive management programs that involve extensive physical and financial coordination.

Organizations and property rights exhibit complex and contradictory relationships with the emergence and practice of conjunctive management. Property rights and organizations are necessary ingredients of conjunctive management, but if they are not properly structured or appropriately positioned they may inhibit conjunctive management.

Preview of Our Comparative Study

In a sense, stating that the physical setting, institutional setting, or organizational setting is important to conjunctive management is stating the obvious. The critical issue is *how* they matter, especially institutions and organizations. Our argument in this chapter has been that organizations and institutions affect not only the likelihood that conjunctive management will be implemented, but also the purpose, form, and operation of conjunctive management projects in relatively complex ways.

To ascertain how institutions and organizations condition conjunctive management in practice, we rely on the considerable variation in the assignment of water rights and organizational forms and responsibilities that exists at the state level in the United States. Other institutional scholars have recognized the usefulness of this variation as a setting for empirical study. As noted by Gardner et al. (1997, *219*):

> State governance of groundwater resources in the western United States provides an institutional setting to study the effects of property rights and regulations ... Independent state authority over groundwater [has] resulted in adoption of four distinct legal doctrines governing groundwater in the 17 western states. Each doctrine established a set of rules directing entry and allocation rules. Further, concern about the pace of groundwater mining has spawned major legal reforms in five states within the last 25 years ... The variety across states of general doctrinal principles and specific regulations creates a diverse set of groundwater property-rights systems in the West.

As water providers and water users in California, Arizona, and Colorado have devised and implemented conjunctive management using similar methods to address similar water problems, they have done so under substantially different institutional conditions concerning water-rights laws and water management organizations. We have studied and compared conjunctive management in the three states to see what difference these institutional and organizational variations make—in the likelihood of adoption and in the manner of successful implementation of conjunctive management. In the next three chapters we provide an in-depth description and analysis of their water institutions and organizations and the practice of conjunctive management.

Part II

How Institutions Matter: Institutions and Conjunctive Management in California, Arizona, and Colorado

4

California

We begin our discussion of the three states with California, where conjunctive management has been practiced in some locations since the 1920s. California's physical circumstances create substantial need, and great opportunities, for conjunctive water management. That combination of need and opportunity establishes a vast potential for the implementation of conjunctive management, but much of that potential remains unrealized despite Californians' long acquaintance with the concept. Reputable organizations readily supply estimates that California could reap a million acre-feet or more per year of additional usable water supply if conjunctive management were practiced in a more coordinated fashion and on a wider scale (Association of Ground Water Agencies 2000; Natural Heritage Institute 1997). That additional supply would close much of the state's chronic long-term supply–demand gap.

California's experience with conjunctive management illustrates how institutions matter in facilitating or inhibiting collective organization and collective action. On one hand, traditions of the state's political system such as local home rule and supporting the creation of special districts and other local governments have promoted the institutional entrepreneurship that is largely responsible for the conjunctive management and many other water projects developed in California over the past 75 years. On the other hand, characteristics of California water-rights law make the development and implementation of every conjunctive management project an arduous process, and impede the wider and better coordinated application of this water management technique for which the state is otherwise so well suited.

The Physical Setting

The water supply and demand characteristics of California make large-scale water storage attractive and feasible. Although it contains substantial expanses of desert, California is not a water-poor state in the same fashion as neighboring Arizona and Nevada.[1] Water supplies and demands in California are, however, extremely variable over space and time.

Precipitation in California, as in much of western North America, is seasonally variable to a degree not experienced in the east. Nearly the entire year's rain and snowfall are concentrated in the months of November through March, with a little occurring in October and April, and almost nothing from May through September. Water demand in California, especially for irrigation but also for municipal uses, is highest during the long and primarily dry summers. Water management efforts in California have always emphasized conserving winter precipitation for summer uses.

Precipitation in California is also greatly variable from year to year.[2] Pacific climatic oscillations affect weather throughout North America, but with its long Pacific-fronting shoreline California is particularly exposed to these yet unpredictable variations. During some years, waves of storms surge in from the ocean across the state, causing floods and landslides and massive erosion of stream banks and beaches. Other years can be extremely dry. Brief and severe droughts may occur, as in 1976–1977 and 2001–2002, but also strings of several consecutive dry years as in 1987–1992. Conserving water and controlling floods in wet years, and saving that water for dry years has also been a recurring focus of California water management.

California's coastal location plays a role in water management in two other ways. Precipitation that is not conserved—that is, not captured in surface reservoirs, diverted to off-channel uses, or sunk underground—runs off into the ocean, from which it cannot be recaptured and used again later.[3] Not only are precipitation and surface runoff along the coast itself lost to the ocean, but most of the rivers that drain California's mountains and interior valleys also empty into the Pacific. Thus, within the limits imposed by flood-control needs, water management in California has emphasized conserving and storing fresh water supplies to keep them from disappearing into the ocean.

In addition, California's coastal location adds a saltwater intrusion threat to near-coast supplies. If flows from the Sacramento and San Joaquin rivers diminish below a certain level, for instance, salt water from the San Francisco Bay begins to creep inland into the channels of the Sacramento–San Joaquin delta, impairing the water quality for freshwater fish species and human water users. Along the Pacific coast and around San Francisco Bay, groundwater basins are vulnerable to the inland movement of salt water when underground water levels are drawn below sea level. Maintaining adequate river flows in the delta and adequate groundwater levels in coastal aquifers are therefore also necessary elements of water management in California.

Perhaps the most easily noticeable feature of California's physical setting with respect to water is its geographic variability. The issues of temporal variability and salinity affect the entire length of the state, but not all areas in the same ways or to the same degree. Most of California's rainfall and snowpack are deposited on the northern half of the state—in the mountains of the north coast, the Sacramento River drainage basin, and the Sierra Nevada range. At more than 40 inches per year, average precipitation in much of northern California is as great as that along the Atlantic seaboard. The southern half of the state, on the other hand, receives substantially less than half that amount: rainfall averages 15 inches per year in Los Angeles, and as little as 5 inches per year in the inland deserts.

Millions of Californians live in northern California, and the northern half of the state is an immensely productive agricultural region, but the majority of Californians and the majority of irrigation water use are found in the southern half of the state. It is often repeated that two thirds of the water is in the northern half of California but two thirds of the people are in the southern half. This geographic mismatch has driven water management in California toward moving water from where it is found to where it is needed.

Water storage—capturing it where and when it is available, and having somewhere to keep it once it is moved closer to the point of use—is a key component of that endeavor. California's surface water reservoirs have a combined capacity of approximately 43 million acre-feet. Even that impressive figure is dwarfed, however, by the state's usable underground water storage capacity, estimated to be 425 million acre-feet (Hauge 1992, *15*).[4] Fortunately for Californians, large groundwater basins underlie some of the areas where precipitation is least plentiful but water is most needed. In the southern half of the state, for instance, the San Joaquin valley, Mojave desert region, and portions of the central and southern coast rest on alluvial soils capable of receiving and retaining large amounts of water. Even in northern California, where average precipitation is greater but seasonal variability and exposure to drought still leave water supplies vulnerable, alluvial aquifers lie beneath the Sacramento River valley, portions of the eastern and southern Bay Area, and the coastal communities of Monterey County.

These elements of California's physical setting shaped the development of water use in the state. Native Californians, and the Spanish missionaries who entered the region in the 1700s, engaged in such low-intensity water use that surface streams and ditches were adequate to convey enough water for agriculture and other consumptive uses. As agricultural development intensified in the 1800s, larger-scale ranchers undertook impoundment projects such as small dams and diversion structures to capture and distribute water. Smaller farmers formed mutual water companies to finance similar projects. Artesian groundwater conditions in some, primarily coastal, areas furthered agricultural development by providing water at the location of use.

Droughts occasionally visited huge losses on California agriculture in the late 1800s, but large-scale efforts to secure water supplies that could withstand California's variability began in earnest in the early 1900s. Those efforts quickly took the form of moving water from places of abundance to places of vulnerability. San Francisco, situated above a largely rocky peninsula surrounded by salt water, needed a more plentiful and more stable water source and constructed a system of canals and tunnels to convey water across the state from the Hetch Hetchy valley of Yosemite. Los Angeles, anticipating that future growth would outstrip its local river and groundwater supplies, built an aqueduct to carry water from the Owens Valley.

In the 1920s, the East Bay Municipal Utility District, serving Oakland and several other communities along the eastern edge of San Francisco Bay, constructed a tunnel system to divert Mokelumne River water from the Sierras to its rapidly growing service area. In the 1930s, the Metropolitan Water District of Southern California, a special district established by the California legislature at the request of a consortium of Southern California cities, built an aqueduct across the desert to the Colorado River.

Also during the 1930s and 1940s, the federal government funded and built the Central Valley Project, which redirected water flows within the San Joaquin valley and conveyed water southward into the valley from northern California and the Sacramento–San Joaquin delta. Finally, from the 1950s through the 1970s, the State of California built the State Water Project, which impounds northern California water and transmits it south, through the delta and via the California Aqueduct to the central and southern coastal regions, the San Joaquin valley, and the southern California deserts.

These large aqueducts traversing California from the north to south and from east to west most obviously helped secure more stable and plentiful sources for the urban and agricultural needs of the coasts and valleys, but they also provided part of the infrastructure for conjunctive water management. The remaining infrastructure came, as it were, from the ground up.

Groundwater use accelerated throughout California even while the big aqueducts were being planned and built. Throughout the twentieth century, farms, industries, and municipalities in California with available groundwater found that the supplies beneath their feet provided relatively inexpensive sources of fresh water that were less erratic than the precipitation and streamflows on which they would otherwise have to depend. In the most intensively developed areas of the state, such as the San Francisco Bay area and the Southern California coastal plain, groundwater pumping exceeded natural replenishment as early as 1920. Evidence of overdraft also appeared in the San Joaquin Valley in the 1920s.

Decades of overdraft followed in several parts of the state, with declining groundwater levels not only raising pumping costs but also triggering more dangerous long-term conditions of land subsidence and seawater intrusion. By the 1960s, when most of the large surface water projects were finished or nearing completion, the California Department of Water Resources identified several groundwater basins that were in various stages of overdraft, up to and including critical levels. Most of those areas also had access to local surface supplies—albeit limited in many locations—and imported water from one or more of the big projects. Physically, therefore, the stage was set for conjunctive management to become a leading water management tool in California—smoothing out the geographic, seasonal, and year-to-year variability of precipitation, and shoring up groundwater basins against the threats of overdraft, subsidence, and seawater intrusion.

In a few of those areas, water users had already organized conjunctive management operations. Others have been initiated since that time. In each location, however, some of the toughest obstacles that had to be overcome in order to implement conjunctive management were the institutional arrangements for allocating and managing water in California.

The Institutional Setting

Water management responsibilities and authority in California are not centered in state government, but distributed between the state and local governments. Within state government, several agencies either have water management responsibilities or have other responsibilities that affect water management. At the local level, water management functions are performed by a large number of governments, including hundreds of counties and municipalities, thousands of special districts, and dozens of joint-powers authorities. Some private groups such as water associations also play important roles.

All of these organizations interact with or represent individual water users, who hold rights to use surface water or groundwater supplies under various provisions of California water law or local court decrees. Those water-rights provisions affect what individual water users are able or willing to do, and thus what water management organizations are able to accomplish. In this section of the chapter, we describe first the elements of California water law that relate to conjunctive management, and then the main types of water management organizations that must or may participate in conjunctive management projects.

California Water-Rights Law

California laws and regulations governing rights to the use of surface water and groundwater bear some similarities to Arizona and Colorado water laws, which are covered in the next two chapters. Other elements are unique to California, and taken as a whole, California water law is one of the most complicated in the United States.

California law recognizes separate bases for establishing rights to the use of groundwater and surface water. Some underground water, however, is treated as if it were surface water for water-rights purposes. Different institutions are relied on for recognizing and determining groundwater and surface water rights. Furthermore, surface water users have more than one basis in California law for claiming a right, and groundwater users have more than one basis in California law for claiming a right.

Existence of these alternative systems for determining rights of use adds considerable uncertainty to water-development and -management efforts in California. Having two different systems for surface and groundwater rights in California, not to mention having complicated systems for each, has not provided incentives for the creation of conjunctive management programs. As noted in the preceding chapter, conjunctive use requires that surface and underground water supplies be treated—in fact if not in law—as more or less interchangeable. California not only separates groundwater and surface water rights, but also defines different rights for different uses and different rights for different types of groundwater.

Surface Water Rights. Rights to the use of surface water resources in California are recognized in two categories: riparian and appropriative. Riparian rights are the rights of landowners to the use of the flow of a stream that crosses or borders their land. These riparian rights are not quantified and thus are not governed by a permit process. Water use by riparian owners is limited only by the doctrine of "reasonable and beneficial use."

The doctrine of prior appropriation is followed with respect to most other uses of surface water resources. Prior appropriation, used also in Arizona and Colorado, awards water rights based on diversion from the stream or water body and actual use, and establishes a priority among rights based on seniority of use. Water may be, and usually is, appropriated by individuals for use on lands that are not adjacent to the body from which the water is diverted.

Prospective surface water appropriators apply to the California Water Resources Control Board for permits to divert and use specific quantities of water per year. Applicants must show a beneficial use for the water, and prove that their diversion and use will not interfere with the rights of senior appropriators or transgress regulatory protections of species, habitat, or other public values. The board also has the authority to determine that a surface water source has been fully appropriated, and on that basis to deny applications for any additional uses of that source, regardless of whether the proposed uses would meet the other criteria on which applications are reviewed (California Water Resources Control Board 1994).

The permit process and the state board were established by the California Water Code, adopted in 1914. Appropriative rights that were recognized before then, and have been exercised continually since, are known as "pre-1914 rights." These use rights are exempt from the permit process and its requirements, and have been treated in the California courts as virtually unalienable and limited only by the state constitutional requirement of reasonable and beneficial use.

Exceptions and Limitations. Four notable exceptions or limitations in California's riparian and appropriative rights systems governing surface water use have had significant implications for conjunctive management in the state.

1. Operation of large surface water projects. First and most important, three of the largest surface water systems in California—the Sacramento, San Joaquin, and Colorado rivers—have been allocated in a substantially different way. Although the legal doctrines of riparian and appropriative water rights still apply formally to the use of water from those rivers, the flows in each river are controlled by large-scale water projects. The operators of those projects deliver water to contractors or member agencies in their service areas, and those contracts or member-agency agreements govern the actual allocation of project water.

The California Department of Water Resources operates the State Water Project, which transmits water from the Sacramento River watershed to contractors in the central and southern portions of the state via the California Aqueduct. The U.S. Bureau of Reclamation operates the federal Central Valley Project, which supplies water along the San Joaquin River to contractors in the central portion of the state. The state's Colorado River allocation is established by an interstate compact and by operating guidelines that govern reservoir levels and releases from Lake Powell and Lake Mead hundreds of miles upstream from the state. The Colorado River water that reaches California and is available for its use is divided among a small number of rights holders, including the Metropolitan Water District of Southern California (MWD). MWD operates the Colorado River Aqueduct, which supplies water to the district's member agencies in southern California under an evolving system of contracts and prices.[5]

The water supplied by those systems comprises much of the surplus water stored in conjunctive-use projects in California. The distribution of the water from those systems—including how reductions will be allocated in dry years and the terms on which additional supplies are made available during wet years—is controlled by contractual provisions and agency policies.

2, 3. Public trust and public nuisance. The second and third limitations on California's system of riparian and appropriative rights are the public trust doctrine and the common law of public nuisance. Both have been employed against the appropriation by the Los Angeles Department of Water and Power (LADWP) of water from the Owens Valley and from streams feeding Mono Lake.

The public trust doctrine may limit diversions and uses of water that threaten harm to publicly held values. In a lawsuit over LADWP's diversions from the streams that feed Mono Lake, the doctrine was applied to find the department liable for the negative effects on Mono Lake.[6] The common law doctrine of public nuisance was raised in a lawsuit over the Owens Valley, charging that LADWP's appropriation of water has caused a variety of harms to the local residents, including adverse health effects resulting from a significant increase in blowing dirt and silt from the dried-up Owens lakebed.[7] Because LADWP used Owens Valley and Mono Lake water to supply its conjunctive management program in the San Fernando Valley, reductions in deliveries resulting from the settlement of these lawsuits have had substantial ramifications for that program.

4. Pueblo water rights. The fourth notable exception to the appropriative rights system is the often-litigated but rarely recognized doctrine of pueblo water rights. Settlements established as pueblos during Spanish colonization were granted under Spanish law a right to use as much water as needed for the residents of the pueblo. The validity of this right was carried over into Mexican law after independence from Spain, and was subsequently recognized by treaty as a valid property claim when California passed to the United States.

A pueblo water right, thus far recognized only for the cities of Los Angeles, San Diego, and San Francisco as successors to their respective pueblos or presidios, is superior to any and all appropriative rights claims, and may even be recognized as dedicating the complete flow of a stream, for example, the Los Angeles River, to the pueblo successor. In the case of the Los Angeles pueblo right, that institutional device was an important element in the development of the LADWP's use of the groundwater basin underlying the San Fernando Valley for storage and recovery of groundwater that was part of the Los Angeles River system.[8]

Groundwater Rights. Rights to the use of underground water supplies in California are recognized and allocated by a multifaceted and sometimes overlapping set of rules that can charitably be called complex. We begin with the legal distinction between underground water that is regarded as surface water and underground water that is regarded as groundwater.

Underground Streamflow. Underground flows of surface water streams are distinguished in California law from "percolating groundwater" or groundwater per se. Underground water *may* be a hydrologically continuous portion of the flow of a stream. In many places in the Southwest, a stream may seem to disappear into the earth, showing just a dry bed for hundreds of feet or even for several miles, but reappearing at the surface farther along. In between, the stream still exists and still moves, but does so underground.

In California water-rights law, underground water that is actually moving in a recognizable "channel" and flowing along a definite path is treated as if it were surface water. An appropriative right to the use of that water may be obtained from the California Water Resources Control Board through the permit application process. The flow of an underground stream may also be included in the "pueblo right" of a pueblo successor, as the water provides part of the base flow of the stream to which the successor has complete rights.

Percolating Groundwater. Unless shown to be underground streamflow moving in a definite channel, in California water extracted from underground is presumed to be percolating groundwater or groundwater per se. The rules regarding allocation of groundwater use rights are quite different from those governing surface waters and underground streams.

Use rights to percolating groundwater are not within the jurisdiction of any state agency, and no statewide procedure exists for the recognition or modification of appropriative rights to the use of groundwater.[9] Instead, rights to the use of groundwater constitute a form of "common law," developed and enforced primarily by courts. Even the threshold question of whether an individual is pumping groundwater per se or underground streamflow—which determines the system of rights one's pumping falls under—is often necessarily decided in court.

California courts have recognized and applied the following possibilities for acquiring and defending rights to use of groundwater (Blomquist 1998; Littleworth and Garner 1995):

1. Overlying landowners have nonquantified rights to pump groundwater for beneficial use on their overlying land. Shortages arising from the "commons" problem this system encourages are allocated according to the doctrine of correlative rights, which means that overlying owners are entitled to a proportion of the aquifer's sustainable yield that corresponds to their proportions of the overlying land.
2. If overlying owners' uses do not exhaust the aquifer's sustainable yield, individuals may appropriate the remaining surplus by pumping and putting it to use. As the surplus diminishes, appropriators may be eliminated in reverse order of seniority.
3. Overlying owners and senior appropriators must be vigilant concerning the presence or absence of surplus groundwater in order to protect their rights. An appropriation of nonsurplus groundwater exercised notoriously and continuously without objection from overlying owners or senior appropriators may ripen into a superior prescriptive right.
4. Any individual or organization that imports water into a watershed for use on the land also has a right to pump and use the return flows of their imports.
5. Pumping rights may also be, and in several cases have been, acquired by adjudication. These quantified rights may derive from a stipulation among the parties, or from a determination by the court based on any combination of the above doctrines.

There is no statewide system for the determination of correlative, appropriative, prescriptive, or return-flow rights. The existence and quantity of these rights in California have been determined on a basin-by-basin basis,[10] when they have been determined at all.

Stored Groundwater, and Underground Storage Capacity. It is especially important to note for purposes of contemplating conjunctive management that these methods determine rights only to the annual yield of a groundwater basin. No system exists in California water law for determining and allocating rights to the storage capacity of a groundwater basin.

Specific court decisions have recognized the rights of appropriators and overlying landowners to recapture the return flows of water applied to the land surface, and a state statute protects groundwater users against any alleged forfeiture of pumping rights if they reduce their pumping as part of an "in-lieu" form of groundwater conservation program, but these protections are only potential components of a framework for recognizing rights to store and recapture groundwater. The general situation with respect to groundwater storage remains as Littleworth and Garner (1995, 52) expressed well:

> To date, regulation of groundwater storage has been left to the courts on a case-by-case basis, and little precedent exists. Many potential issues have yet to be addressed and remain unresolved; for example: Must space be reserved for the normal water level fluctuations during wet and dry cycles? Do overlying entities have a priority over storage and use of water for distant areas if storage capacity is limited? Can local overlying water districts charge for the use of storage capacity underlying their areas? Can counties or other local agencies regulate the use of underground storage capacity? What happens if the imported water degrades the quality of the native supply? . . . More regulation by the courts, the state legislature, or local government can be expected.

In California, the typical case, that is, not in adjudicated basins, is that water in underground storage is treated as a common resource available to all overlying landowners, with no specific user rights assigned. Water in surface water storage is treated differently, such that releases are allocated by surface water users in accordance with their appropriative rights, which are user-specific, quantified, and limited (Natural Heritage Institute 1997, 3). California's different treatment of water stored above ground and water stored below ground presents a further complication to conjunctive management, which would coordinate the two storage resources.

Water Management Organizations

As mentioned earlier, California has not approached water management as a state government function. Consistent with the state's political tradition of supporting local governments and home rule, the California state government has operated mainly to support local water management. The state government has performed this support function by acceding to most requests for the creation of local special-purpose districts or agencies, by providing information on water conditions, and by operating the State Water Project.

The California Water Resources Control Board and the California Department of Water Resources are the two prominent state agencies concerned with the allocation and management of water supplies. The Water Resources Control Board administers the surface water right permit process, but also includes a system of Regional Water Quality Control Boards with authority to issue rulings, orders, and fines concerning land or water uses that may impair water quality. The Department of Water Resources operates the State Water Project and conducts studies of water conditions and the hydrogeologic properties of water resources.

All other water supply management organizations in California are local. California contains an immense number of water districts and other local water supply agencies—2,850 according to a recent count.[11] An alphabetical listing of these published periodically by the Department of Water Resources runs to dozens of pages.

Most water districts in California were created under general-purpose enabling acts, which are numerous. Each act creates a class of water districts with a different mix of authority and responsibilities from the districts created under some other act. These enabling acts have been the basis for county water districts, irrigation districts, California water districts, municipal water districts, flood control and water conservation districts, water storage districts, and community service districts. Some of these agencies have powers that would be compatible with developing and implementing conjunctive management programs; others do not.

California also features many special-act districts, created by their own legislation. Examples include the Metropolitan Water District of Southern California, the San Diego County Water Authority, the Orange County Water District, the Antelope Valley–East Kern Water Agency, and the Mojave Water Agency. Most of these were created to authorize surface water projects or to establish local or regional contracting agencies for one of the large-scale water projects in the state. Some of them have groundwater management powers that could be useful in a conjunctive-use operation; others do not.

Many municipalities have their own water utilities, some of which become extensively engaged in water management activities as well, for example, the Los Angeles Department of Water and Power, San Francisco Water Department, and San Diego Utilities. A number of these were established to develop surface water supplies for municipal distribution, although others have relied on groundwater or a mix of both surface and underground supplies. Private water companies still serve some communities, although private water purveyors are less numerous than in the heyday of mutual water companies organized to construct small-scale projects to supply water for irrigation uses.

Although some local water agencies—most notably water storage districts, water replenishment districts, and groundwater management agencies—were authorized to manage groundwater extractions or to develop and operate water replenishment programs, they represent a minority of California water organizations. Most local agencies lack clear authority to manage either groundwater pumping or groundwater storage. Because most types of conjunctive management operations involve or require some control of ground-water storage and/or groundwater extractions, the institutional structure of California water management organizations has not facilitated the development and implementation of conjunctive management programs.

In 1992, in an effort to promote groundwater management within the state while retaining the tradition of local control, the California legislature enacted Assembly Bill (AB) 3030. Under AB 3030, the authority to engage in a wide variety of groundwater management activities[12] is conferred on any type of local water district, as long as it undertakes an extensive process of consultation and planning with all other water agencies and general-purpose governments overlying a basin.

AB 3030 is understood to have opened a door to the development of conjunctive management programs almost anywhere in the state, as long as one or more local water districts take the lead in initiating and completing the consensus-building and plan-development processes created by the law. However, after a decade since the passage of AB 3030, many locations where the process has been undertaken remain in the plan-development stages.

In the meantime, California's tradition of local control has stimulated a new generation of institutional impediments in the form of "area-of-origin" protections. As the Natural Heritage Institute (1997, 4) observed, "Local concerns over the possible local impacts of conjunctive use programs represents [*sic*] an additional problem to be overcome before a maximal scale conjunctive use program can become a reality." These local concerns are currently manifesting themselves in protectionist ordinances being adopted throughout the state, mostly by rural counties. These area-of-origin ordinances would bar the extraction of groundwater supplies from within the jurisdiction for uses outside the jurisdiction. California courts have recognized such ordinances as valid exercises of local governments' "police powers" to protect the health and well-being of their residents.[13]

Understandable constituent pressures have led local officials to adopt such ordinances. Nevertheless, these protections pose significant obstacles to the development of conjunctive water management programs. If enforced, such ordinances have the effect of walling off considerable amounts of the underground storage capacity or surplus surface water supplies in the state from potential use or participation in conjunctive management. For instance, California Department of Water Resources personnel have hoped to establish conjunctive management programs in the Sacramento Valley to facilitate revised operation of the State Water Project. Changes in State Project operation are needed to maintain greater dry-year yields to support environmental needs downstream in the San Francisco Bay–San Joaquin Delta region, while still supplying water for agricultural uses in the Central Valley and urban uses in the Bay Area and Southern California. To sustain those changes, surplus State Project water could be stored underground in the Sacramento Valley, to be recovered in drier period. That recovered groundwater, called on later to supplement Project deliveries, would necessarily be "exported" from the county in which it was stored, which is precisely what the local ordinances forbid.

The Practice of Conjunctive Management

Conjunctive management in California presents something of a paradox. On the one hand, it has been practiced successfully within the state for a long time, with some cases well known among water management professionals worldwide. On the other hand, conjunctive management is used in only a few localities and on a small scale in California compared with the number of sites and quantities of water potentially available, and getting a conjunctive management project underway anywhere in the state takes years, if not decades. California's favorable physical conditions for conjunctive management, and its complex institutional arrangements, combine to produce this paradox. In this section, we discuss the California findings from our three-state study, describe some common characteristics of how conjunctive management is implemented in the state, and relate those characteristics to the institutional arrangements.

Conjunctive management programs were being implemented in only 12 of 70 basins we sampled in California (see the Appendix concerning the sampling procedure). Programs were being planned or actively considered in another 4 of the 70 basins. Table 4-1 summarizes the conjunctive management projects occurring in our California sample, including (1) the number of basins sampled in each hydrologic region, (2) the number of sample basins with conjunctive management projects, (3) the number of projects operating in the sample basins, and (4) the average number of acre-feet of water per year going into the projects.

The 12-of-70 ratio may understate the actual significance of conjunctive use in California, however. Many of California's 400 identified groundwater basins underlie relatively or completely undeveloped areas of the state, including portions of the Sierra Nevada Mountains and the Mojave Desert. Our random sample of 70 basins included 23 that fit this definition of being remote and relatively or completely undeveloped. It may therefore be more appropriate to state that we found conjunctive management operations in 12 of the 47 developed groundwater basins in our sample, and being planned or actively considered in another 4.[14]

Certain characteristics stand out about the 12 basins for which we collected data on conjunctive management projects. Every one of the 12 has some form of locally initiated basinwide governance institution, such as a special water district or a basin adjudication. We found no conjunctive management projects in basins that lacked basin-scale arrangements for governing groundwater use.

Seven of the twelve California basins with conjunctive management projects we found and studied were in intensively developed, high water demand

Table 4-1. Conjunctive Management Activities in a Sample of 70 California Groundwater Basins, by Region

Hydrologic region (and counties)	No. of basins in sample	No. of sample basins with CWM projects	No. of CWM projects identified in sample	Estimated acre-feet of water per year in CWM projects
San Francisco area	8	1	1	5,500
Central Coast	8	2	4	164,050
South Coast (LA, Ventura, Orange, SD)	13	5	17	613,900
Sacramento area	10	1	1	9,600
San Joaquin Valley / Tulare Lake	10	3	9	468,500
South Lahontan (Mojave, Mono, San Bernardino)	13	2	1	0[a]
Colorado Desert (San Bernardino, Riverside, Imperial)	8	1	1	3,700
Total	70	15	34	1,265,250[b]

a. New project.
b. Data missing on projects in three basins.

agricultural areas with histories of severe groundwater overdraft, but with access to controlled or imported surface water supplies. These seven were:

- Antelope Valley Basin
- Lower Mojave River Basin
- Modesto Basin
- Salinas Valley Basin
- Santa Maria River Valley Basin
- Suisun–Fairfield Valley Basin
- Tulare Lake Basin

Eight basins with conjunctive management programs were in intensively developed, high water demand urban or urbanizing areas overlying relatively large groundwater aquifers. These were:[15]

- Antelope Valley Basin
- Modesto Basin

- Orange County Coastal Plain
- Santa Margarita River Basin
- Suisun–Fairfield Valley Basin
- Sacramento Valley, Carmichael area
- Ventura Central Basin
- Warren Valley Basin

In most of the basins, multiple methods of conjunctive management were being employed, and multiple purposes were being pursued. Table 4-2 summarizes information on the duration, magnitude, methods, and purposes of conjunctive management activities in the 12 basins. Conjunctive management programs ranged in age from brand-new to nearly a century old, and in size from a few thousand acre-feet per year to hundreds of thousands of acre-feet per year.

Within those 12 basins were 34 active conjunctive management projects on which we collected information. A basin might include, for example, a direct-recharge project, plus an in-lieu program of surface water–groundwater exchange, plus a coastal barrier to control seawater intrusion. Nineteen projects were operated by a single organization, thirteen involved two to four organizations, and the remaining two involved five or more organizations. The number of participants in a conjunctive management project was only weakly related, however, to the amount of water stored in the project. In other words, the larger the conjunctive management project, the more likely it was to have multiple participating organizations. Projects annually storing less than 10,000 acre-feet of water average 1.6 participants. Projects annually storing more than 10,000 acre-feet of water average 2.2 participants.

Diverse conjunctive management methods are used in California, as would be expected given the state's approach of letting local water users develop and tailor their management activities. Table 4-3 summarizes the methods used in the projects in our sample. In-lieu projects were common, especially in agricultural areas, a pattern we found also in Arizona (see Chapter 5). As noted in Chapter 2, in-lieu projects can be less expensive to start and maintain, as they do not require as many physical facilities as direct-recharge projects.

Direct-recharge methods, used in about 35 percent of the projects, were more common in urban areas. Most of them were in southern California's metropolitan areas and coastal communities. The use of direct recharge was also closely associated with the scale of conjunctive management activity in a basin; direct-recharge operations were common among the basins with larger amounts of water stored. Direct recharge may also be more common in urbanized areas because it provides an opportunity for water reuse, including the use of a blend of treated wastewater and imported or local surface water for recharge.

Return-flow recharge[16] was less common, practiced in about 15 percent of the projects. It was used by irrigation districts or large-scale surface water suppliers. The benefit of return-flow recharge is that it requires little or no addi-

Table 4-2. Characteristics of California Conjunctive Management Programs

Basin	County	Year began	Average acre-feet per year	Method(s) used	Purpose(s)
Antelope Valley	Los Angeles	1976	23,047	In-lieu, direct recharge	Overdraft recovery, seasonal peaking, drought protection
Carmichael area, Sacramento River valley	Sacramento	1908	6,000	In-lieu	Seasonal peaking, drought protection
Lower Mojave River	San Bernardino	2000	n.a.	Direct recharge	Overdraft recovery
Modesto	Stanislaus	1925	50,000	In-lieu	Overdraft recovery, seasonal peaking, drought protection
Orange County Coastal Plain	Orange	1933	275,000	Direct recharge, in-lieu, plus coastal barrier	Overdraft recovery, seasonal peaking, drought protection
Salinas Valley	Monterey	1956	120,000	Direct recharge, in-lieu, plus coastal barrier	Overdraft recovery, seasonal peaking, drought protection
Santa Margarita River	Riverside, San Diego	1945	14,500	Direct recharge	Assure downstream flow, overdraft recovery
Santa Maria River	Santa Barbara	1963	34,100	Direct recharge, return flow	Seasonal peaking
Suisun–Fairfield Valley	Solano	1959	5,000	Lowering water table to accommodate surface deliveries	Maintain water table for agricultural production
Tulare Lake	Kings	1960	10,000	In-lieu, direct recharge	Seasonal peaking, drought protection
Ventura Central	Ventura	1955	120,000	Direct recharge, in-lieu, plus coastal barrier	Overdraft recovery, seasonal peaking, drought protection

Table 4-3. California Conjunctive Management Projects, by Type and Water Use

Project type	Frequency	Percentage of total projects	Av. volume of water per year (in acre-feet)
In-lieu groundwater savings	14	41	479,250[a]
Direct recharge (spreading basins or injection wells)	12	35	400,400
Return-flow recharge (excess irrigation or controlled dam releases)	5	15	377,100
Draw-down of groundwater table to allow for surface water irrigation	2	6	5,500+[b]
Groundwater pumping for surface water supplements	1	3	3,000
Sample basin totals	34	100	1,265,250

a. Missing data on two projects.

b. Missing data on one project.

tional infrastructure to store water underground where it is feasible to use. Return-flow projects take advantage of the portion of water diversions or reservoir releases that, after being used for irrigation, returns naturally underground through the soil. Irrigators or water agencies using this method then recover percolated water through pumping when surface flows are scarce.

As this review of methods suggests, a variety of sources of water were used for groundwater recharge in the projects in our California sample. Five of the projects combined multiple sources of water for basin recharge.

Fifteen projects use imported water from one of the state's major water projects, including the Metropolitan Water District of Southern California's Colorado River aqueduct, the State Water Project operated by the California Department of Water Resources, and the Central Valley Project managed by the U.S. Bureau of Reclamation. Water from these three projects represents the largest recharge water source in our data, about 41 percent of all the water used for conjunctive management.

Although major surface water projects provide the bulk of recharge water in the California projects we studied, the agencies that manage these facilities—the Metropolitan Water District, the California Department of Water Resources, and the U.S. Bureau of Reclamation—generally do not operate the conjunctive management projects. They are involved directly in only 3 of the 34 projects on which we collected data. More often, these large suppliers sell water to local agencies that use it for conjunctive management.[17] Fourteen projects use recharge water from local rivers and streams that are native to project basins. Five projects also use treated effluent or storm water. Four projects rely primarily on groundwater for supplemental pumping or conservation.

The diversity of conjunctive management methods at work in California reflects a tailoring to local circumstances, which is consistent with the state's basin-by-basin decentralized approach. California's locally driven conjunctive management projects are highly basin specific, with few or no cross-locality arrangements for storing or sharing water supplies. Still, a couple of additional generalizations can be drawn about the manner in which conjunctive management projects occur in the state.

Because of the composition of water management organizations in California, almost any large-scale conjunctive management project in the state will involve multiple entities. In most cases, at least one of the agencies that manage the major surface or imported water projects in the state, that is, the California Department of Water Resources, the U.S. Bureau of Reclamation, the Metropolitan Water District of Southern California, and/or the Los Angeles Department of Water and Power, will be involved in supplying surplus surface water when it is available. One or more other agencies that contract for and receive deliveries from that project, for example, State Water Project contractors, Central Valley Project contractors, or MWD member agencies, will be involved, as only they have legal rights to receive the water. One or more— usually many more—of the "retail-level" local water suppliers or users will be involved, as they are the ones who must adjust their operations to employ surface water or groundwater supplies, as well as cooperate in any financial arrangements involved with those adjustments. Also, where applicable, the local government unit or units responsible for managing the groundwater basin itself, for example, a water storage district, water replenishment district, or a special-act district charged with basin management, will be involved.

This multiorganizational approach has provided some benefits in California, operating like an industry with wholesalers, retailers, and multiple incentives to develop the "best deal" for one's constituents or customers (U.S. Advisory Commission on Intergovernmental Relations 1991). On the other hand, there is no gainsaying that it imposes substantial transaction costs on the development and implementation of conjunctive water management in the state.

Those transaction costs appear to have been overcome most often when contractual arrangements for the distribution of project water provide project operators with incentives to encourage conjunctive management by their contractors. As in Arizona, the allocation of water from the major surface water projects in California has important implications for conjunctive management. Each system operator—the state, the Bureau, and Metropolitan— has an incentive to sell surplus water in wet years, to maintain revenues and to free up system capacity in case the next season is wet, too. The contractual relationships with local entities also mean that each system operator has an incentive to try to find ways of making good on water delivery commitments during dry years.

The combination of these incentives has encouraged each large-scale system operator in California to explore possibilities for placing surplus water into storage, and recapturing it during shortages. In several cases in California, these explorations have resulted in additional contracts to store water underground in areas that have available storage capacity. Often these are areas that have experienced groundwater overdrafts in the past.

The Metropolitan Water District of Southern California, which receives Colorado River water from its own aqueduct and northern California water from the state's California Aqueduct, has tried to develop several such conjunctive-use arrangements. For instance, during a series of wet years in the Colorado River basin during the mid-1980s, Metropolitan had access to much more than its normal allotment of water from the river. Metropolitan developed contracts with Coachella Valley Water District and the Desert Water Agency, two local water districts along the path of Metropolitan's Colorado River Aqueduct. Under these agreements, Metropolitan delivered several thousand acre-feet of Colorado River water to the two districts, which stored the water underground to replenish overdrawn aquifers. The agreements allowed Metropolitan to draw on those stored underground supplies in a drought.

In 1992, surplus northern California water was available to Metropolitan via the California Aqueduct. Metropolitan executed an agreement with two water districts in Kern County—the Kern County Water Agency and Semitropic Water Storage District—located along the aqueduct's course to southern California. These agreements allowed Metropolitan to store up to 500,000 acre-feet of its 1992 surplus underground. Metropolitan agreed to pay $80 per acre-foot to store the water, and another $70 per acre-foot later to pump the water back out when needed.

In early 1998, Metropolitan completed another water storage agreement with Calleguas Municipal Water District in Ventura County in southern California. In 2001, Metropolitan entered into an agreement with the Orange County Water District to coordinate additional water storage in that basin, which is already managed conjunctively. Implementation of the agreements with Semitropic, Calleguas, and Orange County will provide Metropolitan with hundreds of thousands of acre-feet of stored groundwater that can be used in dry years to ensure supplies to Metropolitan member agencies.

Finally, it is clear that conjunctive management has developed in some basins as a response to basin adjudications or other restraints on groundwater pumping to address overdraft problems (Blomquist 1992). Conjunctive management represents an effort to increase the yield of an aquifer and restore underground water levels. These can be especially appealing alternatives to trying to arrest overdraft solely by cutting pumping back to native safe yield. As in any political conflict over the distribution of a valued resource, alternatives that "enlarge the pie" are preferable to smaller slices for everyone.

Conclusion

Relying solely on local institutions for governing groundwater and for organizing conjunctive management projects may have limited the number of locations with conjunctive management programs in California. Nevertheless, our data show that local arrangements are capable of producing extensive and enduring conjunctive management programs. To state it another way, California's basin-by-basin decentralized approach may make it harder to initiate conjunctive management projects, but it does not inhibit the size or longevity of those that emerge. Illustrating this point, the total quantity of water used for conjunctive management just in the California basins in our sample nearly matched the entire amount stored in Arizona's more comprehensive state-directed program, and our California sample does not include all of the basins in the state with conjunctive management programs.

The institutional arrangements governing water-resource management in California have probably discouraged smaller-scale projects in areas of lower water demand, but have allowed conjunctive management to occur where the stakes have been high enough for organizational participants to sustain the endeavor. In basins with high water demands and large-scale project facilities, the benefits from conjunctive management, or the potential losses from failing to implement conjunctive management, appear to be large enough to offset the substantial costs of assembling and maintaining the interorganizational coordination needed to accomplish conjunctive management in California.

California's institutional arrangements provide protection of multiple interests, and virtually ensure that conjunctive management programs will not be implemented unless those multiple interests have been accommodated. The benefits of such an approach can be substantial, and include noteworthy long-term stability for conjunctive management programs once they are implemented. Some of the conjunctive-use operations we have studied are now 60 or more years old and continuing to operate in much the same fashion as when they were initiated. The stability of these programs has in many instances helped to keep their operating costs low and economic efficiency high.

Those institutional arrangements also, however, contribute to a project-by-project, deal-by-deal approach to the development and implementation of conjunctive-use programs. The number of parties involved tends to be larger than would be the case in a state with a more centralized water policy and management structure, and almost every party to a potential deal wields a veto. Accordingly, the organizational and administrative costs associated with the development and initial implementation of conjunctive-use programs are very high in California.

The opportunity costs imposed by these arrangements may also be substantial. The barriers to initiating conjunctive-use programs contribute to the underutilization of many of California's hydrogeologically well suited basins, over-reliance on surface storage facilities with their high financial and environmental costs, and avoidable overdrafting of groundwater supplies in several areas which can produce lasting negative effects on underground storage capacity, water quality, and environmental values. California's project-by-project approach may keep it from realizing the more optimistic forecasts of conjunctive management's potential to narrow the state's overall water deficit (e.g., Natural Heritage Institute 1997).

5

Arizona

Arizona is a relative latecomer to conjunctive management. Unlike California localities, which began experimenting with conjunctive management in the 1920s, and Colorado farmers, who began engaging in conjunctive management in the 1970s, Arizonans did not actively use conjunctive management until the 1990s. During the two prior decades, Arizona water users were embroiled in numerous conflicts that produced substantial changes in water laws and regulations, setting the stage for conjunctive management.

This chapter begins with an examination of the raw materials, the water and water infrastructure, with which Arizonans work in pursuing conjunctive management, followed by an explanation of the laws and regulations governing the allocation and use of water. An in-depth examination of conjunctive management projects is built on this physical and institutional foundation. Arizonans have created and followed a conjunctive management path distinct from those of California and Colorado that will present unique challenges and opportunities for the citizens of the state in the near future.

The Physical Setting

Arizona is one of the most rapidly growing states in the United States. Over the last decade, its population grew by 40 percent, which is rivaled only by its neighbor to the north, Nevada, which grew by 66 percent (U.S. Census Bureau 2001). Much of that growth occurred in and around the metropolitan areas of Phoenix and Tucson, which are located in the more arid regions of an arid state. Even though urban areas are rapidly expanding, agriculture remains vibrant. With hundreds of thousands of acres planted in cotton, alfalfa, wheat, and barley, agriculture consumed nearly 80 percent of Arizona's water in the 1990s, compared to 16 percent from municipal and industrial use, and about 4 percent from power and mines (Arizona Department of Water Resources 1999a). As in the case of urban areas, agriculture tends to be concentrated in the most arid regions of the state.

Arizonans do not meet their growing demand for water through rainfall, which varies from 4 inches in the southwest to more than 30 inches in northwestern mountain ranges, although rainfall is vital for naturally recharging groundwater basins. In the most populous areas of the state, annual rainfall averages 12 inches (Western Regional Climate Center 2001). Because of limited rainfall, native surface waters such as the Salt, the Verde, the Agua Fria, the Santa Cruz, or the Colorado are in reality more streams than rivers. Water demands have been met instead by building large surface water storage and distribution systems and by tapping deep and productive groundwater basins, the two basic building blocks for conjunctive management.

Beginning with the 1911 completion of the Roosevelt Dam on the Salt River and the formation of the Salt River Project (SRP), farmers, and later, cities, have relied on project water to satisfy water needs. Today, the Salt River Project includes eight dams on the Salt, Agua Fria, and Verde Rivers, six reservoirs, and 131 miles of canals. It delivers more than 1 million acre-feet of water annually to a 240,000-acre service area that encompasses a large portion of the Phoenix metropolitan area (SRP 2001).

In 1992, the Bureau of Reclamation completed the Central Arizona Project (CAP). The CAP, a 330-plus–mile canal, extends from Lake Havasu in the northwestern corner of the state to Phoenix and Tucson. It is capable of delivering 1.5 million acre-feet of Colorado River water. Managed by the Central Arizona Water Conservation District (CAWCD), its annual deliveries exceeded 1 million acre-feet in 1996 (Central Arizona Water Conservation District 2001).

The other major source of water for Arizonans is groundwater. Underlying central and southern Arizona are large, deep basins made of sands and gravels. The aquifers underlying Tucson are estimated to extend several thousand feet below ground and hold tens of millions of acre-feet of water (Water Resources Research Center 1999). The introduction of turbine pumps after World War II greatly expanded the use of groundwater. Between 1940 and 1953, groundwater pumping for irrigation increased from an estimated 1.5 million acre-feet to 4.8 million acre-feet per year (Mann 1963). Groundwater now provides about 40 percent, or 2,724,000 acre-feet, of the state's consumptive water use (ADWR 2001b). The increased reliance on groundwater in the Tucson area has led to the depletion of approximately 11 percent of the area's groundwater supplies since 1940 (Water Resources Research Center 1999).

The Arizona water setting appears ideally suited for the widespread use of conjunctive management. Large surface water projects and distribution systems provide sources of surplus water that can be moved throughout the systems. Furthermore, the surface water projects are located in close proximity to large groundwater basins with ample storage capacity. Finally, surface water

projects and groundwater basins together are located in the most heavily populated and among the most agriculture-intensive areas of the state. Thus, the
increasing demand for water could be met, at least in part, through well developed conjunctive management programs.

The Institutional Setting

Although the physical environment appears conducive to conjunctive management, the institutional arrangements governing water initially presented substantial barriers to the coordinated use of ground and surface water. In addressing major water disputes, the state legislature adopted a series of laws and
regulations, and created several public organizations to implement these policies, which support and encourage conjunctive management.

Surface Water Rights

As is the case in most western states, native surface water is governed by the
prior appropriation doctrine and administered by a permitting system through
the Arizona Department of Water Resources (ADWR). Most surface waters,
however, have been substantially developed through water projects, such as the
SRP and the CAP, leading the ADWR to declare that "waters of the state are
fully developed" (ADWR 2001a). Therefore, native surface water and rights to
that water have not played a direct and immediate role in conjunctive management in Arizona.[1]

The major source of surface water in Arizona is project water. Project water
is governed by a different set of rules and regulations than native surface water.
Project water is owned and managed by the entities that have built, own, and
manage the water storage and distribution projects.

For the two major projects in Arizona, the managing organizations are the
Central Arizona Water Conservation District and the Salt River Project. The
Central Arizona Water Conservation District (CAWCD), created by the state
legislature in 1971, operates the CAP. The District possesses the authority to
levy ad valorem taxes on property within its boundaries. The taxes are used for
operations, repayment to the federal government for construction costs of the
CAP, and water storage (CAWCD 1999, 25). The CAWCD has entered into
long-term water contracts, typically 50 years in duration, with municipalities,
industries, and Native American tribes. The water covered by the contracts is
nontransferable—a CAP contractor cannot sell or lease that portion of its
water that it does not use. Each fall, contractors place water orders with the
CAWCD for the following year. All contractors are required to pay operation
and maintenance fees on all water scheduled for delivery (CAWDC 1999, 33).

In 2003, CAWCD charged long-term subcontractors $66 an acre-foot, which includes maintenance fees and energy costs for delivery (CAWCD 2003, rates). Contractors are generally free to use their CAP water as they see fit, including storing it underground.

Initially, the CAWCD signed long-term contracts with irrigation districts, which are the major users of CAP water. However, once the costs of CAP water, the operations and maintenance fees, energy fees, and other charges were specified, irrigation districts indicated that they would use very little CAP water, relying on groundwater instead, and that they might default on their contracts. Rather than lose their agricultural customer base, which would have had severe implications for CAWCD's repayment obligations to the federal government, the CAWCD negotiated a new set of contracts with irrigation districts. Districts waived all or a portion of their original contracts for water and in turn the CAWCD waived all agricultural operations and maintenance fees. Instead, the CAWCD signed 10-year contracts for specially priced pools of agricultural water ranging from $26 to $38 an acre-foot of water in 2003, subject to availability. Municipalities, industries, Native American tribes, and irrigation districts that did not waive their original rights in CAP have priority to receive their water first (CAWCD 2003).

The Salt River Project, storing Salt River and Verde River waters, delivers on average about 1 million acre-feet per year to shareholders and contract holders in its district (SRP 2001), and its water allocations are managed somewhat differently than CAP contracts. In general, SRP has enough direct flow from the Salt and Verde Rivers each year to supply shareholders, or landowners who have senior water rights in its district (SRP 1999). This direct flow is allocated according to the prior appropriation doctrine.[2] SRP supplies its remaining shareholders and contractual users with allocations of stored surface water and pumped groundwater (SRP 1999). Contractual users of the Salt River Project include most of the large water purveyors in Maricopa County such as the cities of Phoenix, Scottsdale, Tempe, Chandler, Peoria, and Mesa. The contracts and water are nontransferable. SRP prepares a Project Reservoir Operations Plan each year to determine its allocations based on reservoir conditions and projected needs of its water users. It adjusts its plan in the spring and fall to account for changing conditions and uses groundwater to supply demands that cannot be met with stored surface water. In general, contractual users have not used large quantities of SRP water for conjunctive management purposes because most contractual users also have CAP contracts, which are available in greater quantities and for incentive prices. SRP itself uses CAP water for conjunctive management by taking excess CAP water from large municipal suppliers with CAP contracts. It then supplies the CAP water to its shareholders in lieu of groundwater.

Project water, not native surface flows, is the primary source of surplus water used for conjunctive management. The predominance of project water means that the organizations that manage it play a central role in conjunctive management. The water management decisions that the SRP and the CAWCD make concerning allocations and pricing directly impact their members' approach to conjunctive management. However, SRP and CAWCD are not only important water managers, but they are also active participants in recharge projects. Thus, it is state-created special water districts that play a central role in conjunctive management.

Groundwater Rights

Before the state's major project operators and other water users began to engage in conjunctive management, Arizona implemented some major changes to its historical system of groundwater rights. Prior to 1980, groundwater in Arizona was governed by the reasonable use doctrine. This doctrine, derived from English common law, treated water percolating through the soil, as opposed to water found in channels, as belonging absolutely to the owners of the soil (Mann 1963, 45). Landholders above a groundwater basin had the right to pump as much water from the basin as they wanted as long as they put it to reasonable use. The reasonable use doctrine does not establish pumping limits. Thus, overlying landholders can pump more water than is naturally replenished, causing water tables to decline, subsidence to occur, and surface vegetation to die. By the mid-twentieth century, Arizona landholders began to experience these types of problems (Mann 1963). The state legislature adopted legislation to address severe groundwater overdraft; however, the legislation did not impose limits on pumping, but only on the spacing, number, and size of wells, and the groundwater overdraft problems continued unabated (Mann 1963).

In 1980 Arizona passed the Arizona Groundwater Management Act, which was an historic piece of legislation. It transformed the open access condition of the most heavily used basins to one of limited access and managed use. Alarm over groundwater depletions, however, was not the only, or necessarily the primary, motivation for the adoption of the 1980 Act. Rapidly growing cities needed additional sources of water. Much of the available groundwater sources in the areas outlying the cities were owned and controlled by farmers. State court decisions made it difficult for cities to acquire and transport such water to their service areas (Leshy and Belanger 1988). Cities were reluctant to buy groundwater from farmers because they felt that farmers were the primary cause of the overdraft problems (Leshy and Belanger 1988). That left the CAP as the most promising source of future water supplies. However, the Secretary of the Interior credibly threatened to eliminate funding for the CAP unless the state moved to strictly regulate groundwater use (Leshy and Belanger 1988). In response to threats to the continuing viability of groundwater basins and to the CAP, representatives of cities, farmers, and mines; state legislators; and the governor negotiated an agreement concerning the management and use of groundwater that the legislature passed in 1980.

The 1980 Arizona Groundwater Management Act established a goal of "safe yield," to be achieved by 2025, for the state's most heavily used groundwater basins, those underlying the Prescott, Phoenix, and Tucson metropolitan areas. Achieving safe yield meant balancing the quantity of groundwater withdrawn with the quantity of water that replenishes an aquifer, both naturally and artificially. In the basin underlying the agricultural area between Phoenix and Tucson, however, the goal was to slow substantially the mining of groundwater. The State designated each of the basins underlying the major municipal regions and the basin underlying the agricultural center of the state as an Active Management Area (AMA), over which the ADWR has administrative authority. All other basins in the state would continue to be governed by the reasonable use doctrine.

The goal of safe yield was to be achieved through a variety of mechanisms. Mine owners and farmers were granted quantified and partially transferable rights to specific quantities of groundwater. Farmers' groundwater rights were conditioned by two factors. First, the rights were subject to conservation requirements—the acre-feet of water per acre that farmers were given would slowly be ratcheted down over time. Second, farmers were not allowed to open new lands to irrigation. Only existing irrigated acreage would be eligible for groundwater use. Under the 1980 act, farmers thus received "grandfathered rights" based on historic use, which annually averaged approximately 4 acre-feet of groundwater per acre. Irrigation grandfathered rights can be converted to non-irrigation grandfathered rights when farmland is transferred to nonagricultural use, but these rights cannot be severed from the land. Mine owners and other industrial users were granted another type of grandfathered rights also based on historic use, which are transferable to other locations.

Cities, towns, private water companies, and water districts were assigned "service area rights" rather than quantified groundwater rights. Service area rights are based on gallons of water per day per customer served, rather than feet of water per acre of land. The per-capita service area right is subject to increasingly strict conservation requirements over the life of the act. The 1980 Act did not immediately place direct limits on the amount of groundwater that could be pumped by municipalities; however, the Act did direct the ADWR to develop rules creating an assured water supply program. The assured water supply program would require municipal and residential water purveyors to demonstrate that they have water of adequate quality and quantity to supply all new and existing uses for 100 years. The goal of the program was to wean municipalities and developers away from mined groundwater. The assured water supply rules, however, were not adopted immediately. The adoption of these, as well as other new institutional rules, in the 1990s set the stage for conjunctive management in Arizona.

Institutional Arrangements Promoting Conjunctive Management

By the mid-1980s, multiple water conflicts emerged in the context of the 1980 Act. The Department of Water Resources struggled to define assured water supply rules that were in accord with the safe yield goal of the Act. Some cities, in anticipation of assured water supply rules, which would almost certainly place strict limits on the use of mined water from managed basins, purchased large ranches overlying basins governed by the reasonable use doctrine just for the water rights. Known as "water farming," cities intended to develop the groundwater on the ranches and transport it to their service areas (Checchio 1988). Municipal purchases of ranches alarmed rural residents who feared that their water would be pumped out from under them. While municipalities searched for additional sources of water in rural areas, developers attempted to stall the adoption of assured water supply rules. Developers were in a precarious water position. Because they were not granted contracts to CAP water they must either tie into a municipal utility or pump groundwater. Assured water supply rules would eliminate groundwater pumping, restricting developers from building out their lands. Finally, water purveyors who wanted to begin experimenting with storing surplus surface water underground realized that state law prevented such activity. State law did not recognize recharged water, and water purveyors feared that they would lose the rights to any water that they stored underground. Over the course of a decade, each of these issues was addressed, and in ways that promoted conjunctive management.

Of these issues, the least controversial was changing state law to recognize groundwater recharge. In 1986, the Arizona legislature passed the Groundwater Storage and Recovery Projects Act, later revised as the Underground Water Storage, Savings and Replenishment Act (see Ariz. Rev. Stat. 45-801 et seq., 2000), which allows private and public entities to store surface water in underground aquifers through direct or in-lieu recharge. Water providers may either apply for permits through the ADWR to establish their own underground storage or in-lieu savings program, or they may apply for permits to store specific amounts of water in others' underground storage projects. The ADWR monitors the quantity of water stored in these projects and the organizations involved receive credits for future recovery of the water stored or saved. For in-lieu groundwater savings projects, the ADWR takes a 5 percent cut from the volume of water delivered when accounting for storage credits. The credits accrued under recharge or in-lieu storage can be transferred to other water providers within the project's AMA.

Shortly after the state recognized direct and in-lieu recharge activities, municipal and rural interests finally settled their differences over the practice of "water farming." In 1991, the legislature adopted a bill that subjected existing "water farms" to close regulation, and forbade the creation of additional water farms (See Arizona Senate Bill 1055 (1991) and Arizona Revised Statutes, Chapter 45, Article 8.1, "Withdrawals of Groundwater for Transportation to

Active Management Areas"). The legislation limited the amount of water that could be transported from a "water farm" to an AMA, imposed progressive transportation fees per acre-foot of water transported, and required cities to pay contributions in lieu of property taxes to the county in which the farm resides. Cities were allowed to pump and transport water from their farms only after they had used most of their CAP allotment. If a city declined a CAP contract, it could not count water from its water farm toward its assured water supply requirement.[3] The legislation signaled municipalities that they would have to meet their water needs through the CAP and not through groundwater transported from rural basins.

The ADWR struggled to design assured water supply rules for the most rapidly expanding water users—municipal and private water utilities. Before the utilities would accede to assured water supply rules, they wanted accessible sources of water to be made available to them. Initially, attempts were made to create regional replenishment or augmentation districts that would develop portfolios of renewable water supplies, largely recharged CAP water, that water purveyors could purchase. These attempts faltered under local politics. Instead, in 1993, the Central Arizona Groundwater Replenishment District was created as a subdivision of the Central Arizona Water Conservation District, which owns and operates the CAP. The Groundwater Replenishment District was given the authority to engage in groundwater recharge projects and contract with water purveyors seeking or possessing an assured water supply certificate to provide them with recharge credits as a means of demonstrating a 100-year supply of water (CAWCD 2003). The district was a critical mechanism for allowing developers access to renewable sources of water for their developments.

Finally, assured water supply rules were established in 1994 (ADWR 2003). Entities applying for a certificate of assured water supply are granted an amount of groundwater that may be used over a 100-year period. The amount granted varies by AMA. Water needs in excess of the groundwater allocation must be met through renewable supplies. Eventually, all demand for water will in some manner be met through renewable supplies. Developers may gain approval for their developments only by demonstrating that they either have a contract with a municipal provider who has been certified or designated of an assured water supply, or that the developer has the ability to replace groundwater that it mines, typically through a contract with the CAGRD (ADWR 2003; CAWCD 2003).

In 1996, the legislature created the Arizona Water Banking Authority (AWBA). The state legislature formed AWBA to contract with Arizona water providers to store Arizona's surplus CAP water in permitted recharge projects. AWBA does this by obtaining water storage permits from the ADWR and then delivers CAP water to recharge sites managed by other water purveyors. The credits that AWBA receives for this storage can be recovered during times of drought and allotted to water suppliers throughout the AMAs. AWBA is also authorized to store water underground for entities in California and Nevada (see Ariz. Rev. Stat. 45 2471, 2000).

Arizona has a clear and deliberate policy of promoting conjunctive management, especially groundwater recharge and banking. By discouraging the practice of water farming and by weaning water utilities away from mined groundwater through certificates of assured water supply, the state has focused water users' attention on other mechanisms for developing additional sources of supply. In providing legal recognition and protection of various forms of groundwater recharge and incentive priced surplus surface water, the State has directly encouraged water purveyors to engage in conjunctive management. Finally, the State itself, through the Water Bank and CAWCD, has become actively engaged in developing and participating in recharge projects.

Conjunctive Management in Arizona

The institutional arrangements devised for managing groundwater established a setting conducive for conjunctive management. Although the state developed permit-granting processes and recognized property rights for recharged groundwater in 1986, planning processes meant that permitted recharge projects did not come on line until 1989. Not until 1990, with the recognition of in-lieu recharge projects, did all four AMAs sport at least one project. With the adoption of assured water supply rules in 1994, the 1996 creation of AWBA, and CAWCD's attractively priced pools of water for recharge, conjunctive management thrived (see Figure 5-1).

Conjunctive Management and the AMAs

Recharge projects and the volume of water stored are unevenly distributed across the four AMAs. The ADWR maintained permits for 42 conjunctive management projects in 1997 and 1998. Of these 42 projects, 30 were located in the Phoenix AMA, which had stored 890,000 acre-feet of water; seven were located in the Tucson AMA, which had stored 84,000 acre-feet of water; four were located in the Pinal AMA, which had stored 616,000 acre feet of water; and one in the Prescott AMA that had stored a total of 14,000 acre feet of water (see Figure 5-1).

The Phoenix AMA is ideally suited for conjunctive management. The basin underlying the Phoenix AMA is deep, with abundant storage capacity. The Salt River, with its sand and gravel beds, provides many favorable sites for direct-recharge projects. Furthermore, the area is crisscrossed with numerous canals delivering CAP and SRP water, and water developed by several irrigation districts. The canals allow large volumes of water to be moved throughout the AMA and they directly connect municipal and agricultural areas.

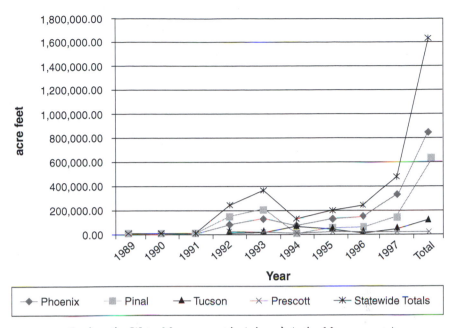

Figure 5-1. Conjunctive Water Management in Arizona's Active Management Areas

In addition to propitious physical circumstances, the demand for and supply of water in the AMA support conjunctive management. The Phoenix AMA encompasses both the greatest demand for water in the state and the largest supplies of project water and effluent, the two primary sources of water used in recharge. In 1995, annual water use totaled about 2.3 million acre-feet in the AMA, with agriculture using about 1.3 million acre-feet, municipalities using about 870,000 acre-feet percent, and industry using about 83,000 acre-feet (ADWR 1999a).

Although the agriculture sector has the highest water demands in the AMA, the percentage of overall water use it represented decreased from 69 percent to 58 percent between 1985 and 1995. Municipal use increased from 28 percent to 38 percent over the same period. As the fastest growing water sector in Arizona, the municipalities within the AMA are actively searching for and developing additional sources of water to meet the increasing demand.

Not surprisingly, the Phoenix AMA is the site of the largest direct and the largest in-lieu recharge projects permitted in the state. The Granite Reef Underground Storage Project (GRUSP), started in 1994, is the largest direct-recharge project in the state, permitted to store up to 200,000 acre-feet of water per year. The Salt River Project holds the permit for the GRUSP and operates and maintains its recharge basins. The Salt River Project coordinates water storage at GRUSP with seven municipalities, CAWCD, and the Water Bank. These entities hold permits to store water at GRUSP for which they receive groundwater credits. They also provide funding for the project, based on each entity's water storage capacity.

GRUSP is located on the bed of the lower Salt River, just below a connection between the CAP Aqueduct and the Salt River canals at the Granite Reef Dam. Thus, it is ideally suited to store both CAP and SRP waters. To date, CAP water has been GRUSP's primary source of water for recharge. Three of the participating cities and the Salt River Project, however, maintain permits to store Salt River water at GRUSP. One city holds a permit to store effluent at GRUSP, but had not delivered effluent to the facility as of 1998. During the first four years of GRUSP's operations water storage totaled nearly 228,000 acre-feet. Since AWBA began participating in the project in 1997, annual storage has increased by as much as 20,000 acre-feet per year. In 2000, the Water Bank alone stored over 80,000 acre-feet of Colorado River water at GRUSP (AWBA 2001).

The Salt River Project also manages the largest in-lieu groundwater savings facility in Arizona. Like GRUSP, this project involves the delivery of water supplies among seven municipal water providers, one private water provider, and CAWCD. These organizations can deliver up to 200,000 acre-feet per year of Colorado River water to SRP, which it uses instead of pumping groundwater. To comply with Arizona's permitting requirements for in-lieu recharge, SRP must demonstrate that it cuts groundwater pumping by an amount that is equivalent to the acre-feet of Colorado River water that it receives from these entities. The water providers involved in the SRP in-lieu project then receive in-lieu recharge credits for their water deliveries, minus the 5 percent cut taken by the State The SRP in-lieu project began in 1996, and by 1997 the Department of Water Resources reported nearly 167,000 acre-feet of groundwater savings for this project.

By volume of water stored, the Pinal AMA is the next most active in conjunctive management. Three of the four recharge projects in the AMA are large in-lieu projects permitted to store from 55,000 to 120,000 acre-feet of water per year. The projects in the Pinal AMA accounted for 40 percent of the total water stored through groundwater recharge in the state. The in-lieu projects center on three large irrigation and drainage districts—the Maricopa-Stanfield, the Hohokam, and the Central Arizona. Historically, the districts relied on groundwater to meet their members' needs. In the 1970s and1980s, the districts signed contracts for CAP water and invested heavily in the required infrastructure to deliver the water to their members. A poor economy, combined with CAP water that was considerably more expensive than groundwater, forced the districts to withdraw from their CAP contracts. The state responded to the crisis in two ways. First, it recognized in-lieu recharge projects. Second, through the CAWCD it made available low-cost pools of CAP water to encourage farmers' use of the water. In 1992 and 1993, CAWCD facilitated the growth of conjunctive management in Pinal County by delivering CAP water to three irrigation districts for in-lieu groundwater savings. In these two years combined, CAWCD received credits for more than 372,000 acre-feet of groundwater savings. In 1994,

CAWCD did not deliver water for in-lieu recharge to the Pinal AMA; however, it did sell CAP water to the irrigation districts at substantially lower prices. The irrigation districts began taking smaller amounts of in-lieu CAP water again from CAWCD in 1995 and 1996, totaling less than 100,000 acre-feet for all three irrigation districts during the two-year period. In 1997, the volume of in-lieu deliveries picked up again when the Water Bank started participating with the three irrigation districts to deliver in-lieu CAP water (ADWR 2001a).

At first glance, it appears surprising that the Tucson AMA houses seven recharge projects that have cumulatively stored only 5 percent of the state's total recharge through the end of 1997. After all, the Tucson metropolitan area is second only to the Phoenix metropolitan area in terms of population and municipal and industrial water demand. Furthermore, Tucson holds the distinction of being the largest city in the United States wholly dependent on groundwater. Moreover, the Tucson groundwater basin suffers from declining water tables, subsidence, and water quality issues. Tucson appears ripe for conjunctive management, particularly as it resides in a physical and institutional setting that strongly encourages it; yet at same time the Tucson AMA has lagged behind its counterparts.

The primary source of water for recharge is CAP water. The Tucson Water Department, the largest water purveyor in the region, signed the single largest CAP contract for almost 149,000 acre-feet of water annually. In 1989, in anticipation of the delivery of CAP water to Tucson, the city council adopted a plan developed by the Tucson Water Department that would involve the treatment and direct delivery of CAP water to the citizens of Tucson. Citizens of the AMA would directly use most of the CAP water delivered and only a small portion of it would be recharged.

The Tucson Water Department faced a daunting task in switching from relatively high quality groundwater to lower quality project water. Unlike their municipal counterparts in the Phoenix metropolitan area who had decades of experience treating and delivering surface water, the Tucson Water Department had never before delivered surface water for consumption. The transition to surface treatment and delivery was less than smooth. In its haste to deliver CAP water, the Tucson Water Department did not replace aging water lines, nor did it test the water, which was more acidic and contained higher levels of dissolved solids than groundwater, to determine its effects on waterlines. Shortly after receiving CAP water in late 1992, many citizens began complaining of discolored water coming from their taps. A number of people also reported broken water pipes and damaged water heaters and evaporative coolers. By 1994, the City Council ordered the water department to cease delivery of CAP water. In 1995, Tucson citizens passed Proposition 200, the Water Consumers Protection Act, which prohibited the direct delivery of CAP water unless its quality was equal to that of groundwater (O'Connell 2001a).[4]

In 1997, Tucson citizens refused to rescind Proposition 200, even though business leaders, developers, and city officials claimed that it was too onerous and would threaten economic development. However, citizens did approve a bond issue for replacing water lines. The Tucson Water Department set about replacing more than 150 miles of water lines and developing a new approach for the use of CAP water (O'Connell 2001b) Conjunctive management is the lynchpin of the new approach. The Tucson Water Department plans to recharge up to 60,000 acre-feet of CAP water each year that will be pumped, blended with groundwater, and delivered to customers (Tobin 2001a). Initially, the blend will consist of 95 percent groundwater; within a decade the blend will consist of half groundwater and half recharged CAP water. In addition, the Tucson Water Department and a number of smaller water purveyors and Indian tribes are actively developing recharge sites throughout the Tucson AMA (Gelt et al. 2001). Thus, within the next decade the amount of water recharged within the Tucson Basin should increase considerably.

The Prescott AMA expands across 485 acres in central Yavapai County and includes two sub-basins (ADWR 2001a). It has only one active conjunctive management project, which is an effluent recharge project managed by the City of Prescott. The limited use of recharge in the Prescott AMA can be partially attributed to its small size, geology, and the location of water providers relative to potential recharge sites (ADWR 2001a). In addition, the municipal, industrial, and agricultural water demands in this AMA are much smaller than in the Phoenix, Tucson and Pinal areas. However, Prescott's municipal growth may pose future problems for the region's groundwater supplies. State law permits Prescott water providers to use groundwater imported from the Big Chino Basin outside the AMA, and pumping from the Big Chino Basin can reduce the available surface water downstream in the growing Verde Valley (ADWR 2001a).

Organizations, Coordination, and Types of Recharge

As of 1998, 31 organizations or political jurisdictions managed conjunctive management projects in Arizona. Municipalities held permits for the largest number of projects, whereas the state special districts, specifically the CAWCD, held the fewest permits. Noticeably absent is the Water Bank. It does not manage projects. However, it does participate by holding long-term storage permits, which allows it to provide water to projects operated by municipalities and irrigation districts. Besides the Water Bank, two cities and three water companies also provide water to projects without managing any of them. Thus, all together, 37 organizations and political jurisdictions participated in the 42 projects permitted in 1997–1998 (see Table 5-1).

The number of permits, however, does not accurately reflect the amount of water held in storage by organizations. Although the CAWCD and AWBA hold the fewest number of permits, as of 1997 they controlled about 70 percent of all long-term storage credits (see Table 5-2). CAWCD held the majority of these

Table 5-1. Arizona Permit Holders, by Organization Type, 1997–1998

Permit holders	Facility permit	Percentage	Storage permit	Percentage
Municipalities	19	45.2	49	48.0
Irrigation districts	11	26.2	0	0.0
Utilities/private water co.	8	19.0	22	21.6
CAWCD	4	9.5	17	16.7
AWBA	0	0.0	14	13.7
Total	42	100.0	102	100.0

credits, although the percentage of credits held by the Water Bank has grown since 1997, when it held approximately 270,000 long-term storage credits from its first two years of operation. Between 1998 and 2000, the Water Bank increased its average annual storage to nearly 300,000 acre-feet (AWBA 2000, 1998a, 1999).

Although the CAWCD develops long-term storage credits for use of its members who contract for them, the state is currently debating how the credits amassed by the Water Bank should be used. The Water Bank is obligated to transfer credits to the CAWCD to meet demands of CAP contractors in times of shortage. In addition to CAP shortages, an Arizona Water Banking Authority Study Commission (1998) recommended that the Water Bank establish policies for a credit loan program to municipal and industrial users, whereby users would either repay with credits or pay the cost of credit replacement. The Study Commission also proposed that the Water Bank be granted the authority to create a credit pool to be used when low-priority CAP allocations are unavailable. Finally, it proposed that the Water Bank's drought protection service be expanded to give credits to water providers during shortages in proportion to the ad valorem taxes paid within the service area experiencing the shortage.

Municipalities hold 27 percent of long-term storage credits. Municipal long-term storage credits constitute a portion of the water portfolios that municipalities have developed to meet the 100-year assured water supply require-

Table 5-2. Arizona's Long-Term Storage Credits, by Organization Type, 1997

Entity type	Long-term storage credits (AF)	Percentage of total credits
CAWCD	796,641	52.7
Municipalities	405,033	26.8
AWBA	269,122	17.8
Utilities/private water co.	41,592	2.7
Irrigation districts	0	0
Total	1,512,388	100.0

ments. Thus, most cities do not anticipate recovering their stored water for several decades.

Clear patterns exist among the 42 permitted projects in our study, in terms of numbers of organizations participating and the type of recharge occurring. Among the 42 projects, just over half involved a single organization managing and storing its own water in the project. Single-organization projects store less water, are more likely to involve direct recharge, and are more likely to use effluent than are multiorganization projects. Of the 22 single-organization projects, 20 involved direct recharge and two involved in-lieu recharge. Also, of the 22 single-organization projects, 14 used effluent as the only source of water stored,[5] five used CAP water only, two used a mix of CAP and other water, and one used surface water other than CAP. The average annual permitted storage among these 22 projects was 3,454 acre-feet.

The organizations typically involved in single-organization projects are municipal or private water utilities that use direct recharge to store water that they own, but that they otherwise would not use. For instance, the town of Surprise, located north of Phoenix, has a CAP contract, but no means of treating and directly delivering the water to its customers. Surprise is required to pay for the CAP water, whether or not it uses it. Thus, by storing water underground, it remains available for use by the town at a later time. Similarly, the Tucson Water Department has developed a series of effluent dominated wetlands. The wetlands further cleanse the effluent before it percolates underground, which is later recovered for turf irrigation in golf courses and parks. The effluent placed in the wetlands would otherwise have been disposed of in the Santa Cruz River bed.

Among the 42 projects we studied, 20 involved between 2 and 10 organizations. Typically in Arizona, a single organization holds the permit for the project and is the project manager while multiple organizations hold storage permits for the project. The multiple-organization projects store more water, are more likely to involve in-lieu recharge, and are more likely to use CAP water than are single-organization projects. Of the 20 multiple-organization projects, 14 involved in-lieu recharge and 6 involved direct recharge. Also, of the 20 multiple-organization projects, 16 used CAP water as the only source of water stored, 3 used a mix of CAP and other water, and 1 used effluent only. The average annual permitted storage among these 20 projects was 49,607 acre-feet.

The typical multiple-organization project involves one or both of the state special districts, one or more large municipal utilities, and an irrigation district.[6] The state special districts and municipal utilities provide the irrigation district with CAP water to use for crop irrigation in lieu of pumping groundwater. In exchange, the state special districts and/or municipal utilities receive long-term storage credits for the groundwater that otherwise would have been pumped. The irrigation district typically receives CAP water at a cost equal to or less than what it would have cost the irrigation district or its members to pump groundwater. The state special districts and the utilities receive long-term storage credits at a cost much less than what it would have cost to create those credits through direct recharge. Also, the utilities are able to put into long-term storage substantial portions of their CAP water allotments, instead of forgoing them until the time when municipal demand would dictate their use.

Arizona water users rely on both in-lieu and direct recharge for long-term storage of surplus surface water. In terms of volume permitted and stored, in-lieu clearly outweighs direct recharge. In-lieu projects, while numbering only 16 in our study, accounted for 70 percent of permitted storage. Also, four of the five largest conjunctive management projects in Arizona, permitted to store between 100,000 acre-feet and 200,000 acre-feet of water annually, were in-lieu projects. In-lieu projects are not only larger, but they are also older. In the late 1990s, the average age of in-lieu projects was nearly 8 years compared with 6.4 years for direct-recharge projects. Thus, although the state recognized direct-recharge projects first in 1986 legislation, and later in-lieu projects in 1990, in-lieu projects form the backbone of conjunctive management in Arizona. Direct-recharge projects, although smaller, also play an important role in conjunctive management in Arizona. Direct-recharge projects are the primary means by which water utilities are able to capture and store surplus surface water that would otherwise be lost to them.

Conclusion

Two factors supported the emergence of conjunctive management in Arizona in the late 1980s. First, the CAP provided an ample source of surplus water. Currently, CAP water constitutes 70 percent of all water stored through direct- and in-lieu recharge projects. However, a plentiful supply of surplus water is not all that is required for conjunctive management to emerge. Conjunctive management requires a supportive set of institutional arrangements. The 1980 Groundwater Management Act provided a foundation on which the state legislature built a series of policies supporting conjunctive management. From 1986, when rights in stored groundwater were first recognized, until 1996 when AWBA was authorized, the legislature created an institutional environment supportive of conjunctive management.

The importance of the institutional setting for promoting conjunctive management is reflected in the goals of recharge projects as identified by the organizations that manage the projects. Thirty-two of the forty-two recharge projects in our study were intended to supplement water supplies through the development of long-term groundwater storage credits. Groundwater storage credits are the primary tool at the disposal of public and private water utilities for meeting assured water supply rules, a requirement of the 1980 Groundwater Management Act. At a distant second is the goal of raising water tables or reducing pumping lifts. Fifteen projects were located so as to address problems associated with declining water tables; of those fifteen, seven included the goal of generating groundwater storage credits. Concern over declining water table levels is also reflective of the 1980 Groundwater Management Act that set the goal of safe yield for the state's actively managed basins.

Conjunctive management in Arizona reflects the interplay between the physical and institutional settings. Deep aquifers with plenty of storage capacity located in areas of great water demand make conjunctive management physically possible. Surplus CAP water makes conjunctive management attractive. However, supportive rules, regulations, and state agencies make conjunctive management in Arizona a reality.

6

Colorado

Colorado shares a number of characteristics with Arizona and California—a rapidly growing population with urban centers and productive agricultural land located in arid areas far from substantial water sources. In Colorado, agriculture and urban life take place in the arid plains and foothills east of the Rocky Mountains. Water to serve agricultural and urban needs is transported hundreds of miles east from major surface water projects built on the water-rich western slopes of the Rockies. As the cost and the contentions surrounding new western slope water projects increase, municipalities and farmers on the eastern slopes are turning their attention toward better coordination of ground and surface water as a means of developing additional water supplies.

Unlike Arizonans and Californians, however, the citizens of Colorado have not developed conjunctive water management in the form of long-term underground storage of water to guard against drought or to address future urban demands. Instead, conjunctive management is employed to allow farmers and municipalities greater access to groundwater, while protecting stream flows from the negative effects of groundwater pumping. Only recently have a few municipalities begun exploring long-term underground storage of surplus surface water.

Conjunctive management occurs differently in Colorado because of a combination of physical circumstances and institutional choices. Colorado's eastern slopes and plains were the first areas developed extensively by settlers from the eastern United States in the latter half of the 1800s. The settlers mined, farmed, and built towns along the banks of streams and rivers. They allocated water and water rights, as well as mining claims, home sites, and homesteads, according to the rule of "first in time, first in right"—the basis of the prior appropriation system discussed in connection with surface water supplies in Arizona and California.

Water demands quickly outstripped the scarce supplies available in arid eastern Colorado. Miners, farmers, and townspeople—and later the state and federal governments—soon began building and investing in canals, ditches, tunnels, pumps, dams, and reservoirs to capture abundant western slope waters for transport to the eastern slopes and plains. The remainder of the first century of Anglo development of Colorado was fueled by the construction of surface water projects.

Not until the 1950s did substantial numbers of farmers turn to wells and groundwater for crop irrigation. Groundwater not only helped protect farmers from droughts, but installing wells and pumps also proved to be much less expensive than building surface water projects.

The emerging use of groundwater touched off decades of conflict between ground and surface water users as groundwater pumping reduced surface water flows. State officials struggled to incorporate groundwater into the prior appropriation system that had been developed to allocate and protect surface water rights.

Conjunctive management has been the mechanism the state has used to reduce conflicts and accommodate farmers' increased use of groundwater. More recently, rapidly growing cities along the eastern slopes of the Rockies—the Front Range—have begun exploring conjunctive management as a means of developing additional water sources.

The Physical Setting

Like most western states, Colorado experienced rapid population growth over the preceding decade. Colorado's population expanded by 30 percent (U.S. Census Bureau 2001). Much of that growth occurred in and around the urban areas of the Front Range, from the Fort Collins area in the north to the Pueblo area in the south. Douglass County, for example, which encompasses many of the communities between Denver and Colorado Springs, grew by 191 percent (U.S. Census Bureau 2001). More than 67 percent of Colorado's population is concentrated in and around the Front Range cities. If one adds the population extending east to the Kansas border, fully 71 percent of Colorado residents live in the eastern third of the state, which is also one of the driest regions of the state.

That eastern third contains much of Colorado's developed agricultural land. The counties of eastern Colorado often lead the state in the production of winter wheat, corn, sorghum, dry beans, sugar beets, and cattle and calves (Colorado Department of Agriculture 2000). As in most western states, agricultural water use in Colorado far outstrips all other types of uses. Irrigation consumes 65 percent of all surface water used in the state per year (Colorado Division of Water Resources 2000a).

Urban and agricultural water demands in eastern Colorado are met by means of a combination of native surface flows, surface water projects, and groundwater. Two major rivers and their tributaries supply a large proportion of those demands. The South Platte River originates in the Rockies southwest of Denver, and flows north and east through Denver and on to Nebraska. Its historic average flow is 408,000 acre-feet per year at the Nebraska border (Colorado Division of Water Resources 2000b). Most of northeastern Colorado lies within the watershed of the South Platte River, including the state's most heavily populated areas and its richest agricultural lands. The Arkansas River also rises in the Rockies southwest of Denver, but flows south and east through Pueblo and on to Kansas. Its historic average flow at the Kansas border is 162,000 acre-feet per year (Colorado Division of Water Resources 2000b).

Both rivers are overappropriated. More water rights have been defined and allocated than can be satisfied in a given year. During the peak demand period of summer on the South Platte River, water rights defined after 1882 are rarely served (Sample 1997, personal communication). On the Arkansas River, irrespective of the time of year, rights in native flows defined after 1887 typically are not met.

In addition to native flows, project water represents an important source of water for eastern Colorado. The Denver Water Board, serving the city and county of Denver, maintains the largest collection of transmountain water projects. From the 1930s, when Denver first began work on the Fraser River collection system in the Colorado River basin, to the present, Denver has built and expanded three major projects that produce several hundred thousand acre-feet of water per year (Denver Water Board 2003).

Two conservancy districts, one in each major river basin, also have built major transmountain projects. The Northern Colorado Water Conservancy District manages the Colorado–Big Thompson (CBT) project, which it built in conjunction with the U.S. Bureau of Reclamation. Water collected from the Colorado River basin is stored in several western slope reservoirs, transported through a 13-mile tunnel to east slope reservoirs, and released as needed to supplement native surface water supplies in the South Platte basin. In a typical year, the project provides 230,000 acre-feet of water to its members, including municipalities such as Boulder, Greeley, and Fort Collins, several irrigation districts extending from Loveland to the Nebraska border, and some ditch companies and farmers. The district's members hold allotments of CBT project water, which they may freely exchange with other members (Northern Colorado Water Conservancy District 2001).

The Southeastern Colorado Water Conservancy District manages the Arkansas-Frying Pan project. Water is diverted from the Frying Pan River in the Colorado River Basin to the Arkansas River via a 6-mile canal. The project serves irrigation districts and numerous municipalities such as Pueblo and Colorado Springs (Abbott 1985).

Several smaller transmountain projects have been built by irrigation districts and municipalities. All together, more than 40 transmountain projects, using 16 tunnels and numerous ditches, transport more than a million acre-feet of water per year from the Colorado River basin to eastern Colorado (Colorado Division of Water Resources 2000c).

Groundwater also supplies eastern Colorado's cities and farms. Farmers have come to rely heavily on tributary groundwater underlying the South Platte and Arkansas rivers. The South Platte basin is estimated to hold 8 million acre-feet of groundwater (MacDonnell 1988), and the Arkansas basin another 2 million acre-feet (Concerning the Amended Rules and Regulations Governing the Diversion and Use of Tributary Groundwater in the Arkansas River Basin, Colorado, Case No. 95CW211, District Court, Water Division 2). Additional deep basins containing sandstone aquifers underlie the Denver metropolitan area. Other groundwater basins are scattered about the eastern plains, largely separate from surface streams, and are used by farmers to irrigate their crops. A portion of the multistate Ogallala Aquifer reaches into this portion of eastern Colorado.

The stages of water development—from native surface water to project water to groundwater—are linked to the development of Colorado's institutional arrangements for allocating and using water supplies.

The Institutional Setting

In terms of institutional arrangements, Colorado differs strikingly from both Arizona and California. In contrast with the situation in Arizona, water development and allocation in Colorado has been largely undertaken and administered at the local level through water districts, groundwater districts, and water courts. To be sure, Colorado uses statewide entities such as the state engineer and the Supreme Court, but they primarily provide administrative support and oversight of activities undertaken at the local level. In contrast with California, Colorado relies on the prior appropriation system to govern rights to both ground and surface water.

Surface Water and Tributary Groundwater

Two words suffice to capture the corpus of surface water law in Colorado: prior appropriation. Beginning in 1876, Article XVI, Section 6 of the Colorado constitution declared that "the right to divert the unappropriated waters of any natural stream to beneficial uses shall never be denied." In 1882, the Colorado Supreme Court held that the prior appropriation doctrine had governed surface water even prior to the adoption of the constitution, and recognized "the doctrine of prior appropriation as the central principle of all water allocation" (Vranesh 1987, 62).

In general, for an appropriation to be recognized and a water right perfected, an individual user must divert water and put it to a beneficial use (Vranesh

1987, *130*).[1] The requirement of beneficial use discourages speculation in water (Vranesh 1987, *141*). Historically, if an appropriator did not use the full amount of water decreed, the portion left unused remained in the stream. Eventually, others could appropriate that water and acquire decreed rights in it (Vranesh 1987, *141*).

The appropriation system separates land ownership from rights to the use of water. Appropriative rights can be transferred, even to water users in a different watershed, and point of diversion and type of use can change, as long as other appropriators are not injured (Vranesh 1987, *120*). The "no injury" rule protects "junior appropriators' rights to stream conditions as they existed at the time the juniors initiated their appropriations" (Vranesh 1987, *72–73*).

The prior appropriation doctrine also governs tributary groundwater— "water which, by its own movement, will become part of a natural surface stream or the use of which will adversely affect the flow of the stream" (Fischer and Ray 1978, *47*). Although Colorado courts had long recognized that tributary groundwater was governed by prior appropriation, most wells had never been adjudicated (MacDonnell 1988, *586*). A 1969 Act of the Colorado legislature designated a special period through June 30, 1972, during which a user of tributary groundwater could come forward and have his or her rights adjudicated with a priority date that would be the same as the true appropriation date. In other words, groundwater pumpers who acted swiftly could retrieve the place in line they would have had if they had joined an earlier adjudication of their stream system. Since July 1, 1972, tributary groundwater rights have been adjudicated and dated in the same manner as surface water rights (Senate Bill 81, "Water Right Determination and Administration Act of 1969").

Administration of the Prior Appropriation Doctrine

What distinguishes the prior appropriation doctrine in Colorado from the same doctrine in other western states is the means by which it is administered. In most states, a state water engineer, state water director, or state water board administers the appropriation system. In Colorado, however, water courts define and enforce appropriative rights.[2]

When an appropriator chooses to seek a decree, he or she files an application with the clerk in the appropriate water court. Once a month, the water court clerks create and publish resumes that list all applicants and the purposes of the applications. Any appropriator may oppose an application by filing a statement of opposition (Vranesh 1987, *442*).

Applications are turned over to water referees. The referees, who are "nonlawyer, technically trained personnel," gather evidence and make initial determinations concerning water rights (Vranesh 1987, *456*). If the issue is relatively simple and free of conflict the referees will often ask the applicants to draft the appropriate decree (Vranesh 1987, *444*).[3] Referees' rulings, if not appealed, are incorporated into decrees by water judges (Vranesh 1987, *445*).

The findings of fact or law of the referees do not bind water judges. If a referee's ruling is appealed, de novo hearings are held before a water judge. Such hearings are more formal than those before the referee, although the rules of civil procedure are not guiding. The courts generally allow the parties to the case an opportunity to propose terms and conditions that would prevent injury, and the judges themselves may suggest such terms and conditions (Vranesh 1987, *446–447*). The Colorado Supreme Court is the exclusive appellate jurisdiction over water cases. Appeals are allowed only with respect to protested matters (Vranesh 1987, *447*).

The system of devising and revising water rights encourages appropriators to negotiate among themselves before bringing their claims to the court. Once before the court, the procedures followed also encourage negotiated settlements. Only issues that cannot be settled among the appropriators go to trial before a judge, and even then the focus is on crafting an agreement acceptable to all parties.

Once a decree is entered, it must be administered, monitored, and enforced. Appropriators, the state engineer, the state engineer's seven division offices, and water commissioners employed by the state engineer participate in monitoring and enforcing water rights. The unit at which most administration occurs is the water district, of which there are 80 in Colorado (Vranesh 1987, *379*). A water commissioner serves each water district. Commissioners measure and report each diversion taken under each water right, and ensure that shutting down junior appropriations satisfies rights of senior water-rights holders. Thus, commissioners carry much of the responsibility for administering and monitoring the prior appropriation system.

Adjudicating and administering water rights at the district level is unwieldy. An appropriator in one district might be junior to an appropriator in another district, even if the date of appropriation was earlier, simply because the adjudication took place at a later time. Also, each district traditionally used a different method for listing priorities, making it difficult to determine one's priority of water use compared to others in different districts (Vranesh 1987, *384*). Eight years after creating water districts and water commissioners in 1879, the Colorado legislature created divisions and division engineers. Water divisions encompass the districts located in each major river basin. Division engineers, who report to the state engineer, are authorized to coordinate water rights across districts. Water commissioners are to follow the direction of the division engineer, such as when the division engineer shuts down diversions in one district to satisfy senior rights in another.

The state and division engineers play the roles of coordinator and information gatherer and disseminator. Division engineers maintain and update lists of appropriation rights and priorities in each division. They determine the accuracy of statements made in water applications and protests. They measure water flows, determine who is in priority, and order junior appropriators shut down. They inspect and monitor diversion works, reservoirs, and dams, ensuring safety and accurate measurement of diversions (Vranesh 1987, *509*). The state and division engineers provide information and technical resources to appropriators, courts, and the state legislature, allowing them to define, revise, administer, monitor, and enforce the water-rights system.

Transmountain Water Projects

Another major source of water in the South Platte and Arkansas River Basins is water diverted and transported from other watersheds, particularly the western slope. Unlike native surface water, transmountain water is not governed by the prior appropriation doctrine. It is owned by the entity that developed and transported it, which can use and reuse it as it sees fit (Vranesh 1987, *345*).

The two major types of entities that have developed such projects are municipalities and conservancy districts. Municipalities, such as Denver, exercise substantial authority in developing water projects. Municipalities can acquire water rights, and hold the authority to condemn land and water rights for such projects. Municipal water departments and districts act primarily as water producers and retailers, developing surface water supplies that are treated and delivered to customers.

Conservancy districts, state authorized special purpose governments created to promote and manage transmountain projects, deal with the federal government in the design and construction of projects, and search for financial support. Conservancy districts serve both agriculture and municipalities, and act as water producers and water wholesalers. Two such districts—the Northern Colorado Water Conservancy District and the Southeastern Colorado Water Conservancy District—were mentioned earlier in the chapter.

Conservancy districts exercise less extensive authority than municipalities. Their eminent domain powers do not include condemning water rights. Furthermore, conservancy districts are required to build compensatory storage into each transmountain water project, an expensive activity that is not required of municipalities. Storage reservoirs are built into transmountain water projects for the express use and benefit of citizens of western Colorado.

Other Types of Groundwater

Under Colorado groundwater law and administration, the "location of the well, the waters which it captures, and the uses to be made of the water all result in different rules" (Fischer and Ray 1978, *47*). One set of rules applies to "designated groundwater basins." Groundwater in designated basins is not part of a natural surface stream system, or moves toward or from a stream so slowly that it is as if it were not part of the stream.

Such basins are governed under the Ground Water Management Act of 1965. That act applied the priority doctrine to uses of groundwater in designated basins, and granted the Colorado Ground Water Commission broad powers to regulate those uses to manage supply and demand. The commission issues permits for wells. Permits specify location, use, acre-feet of water, and gallons per minute of pumping. Permits are granted only if "the proposed appropriation would not unreasonably impair vested rights and that the appropriation would not create unreasonable waste" (Vranesh 1987, 256). In times of substantial shortage, the commission may shut down junior wells so that senior wells can continue to appropriate water (Fischer and Ray 1978, 57).

The recognition of seniority in designated basins does not translate into a full-blown prior appropriation system like that which applies to surface and groundwater elsewhere in the state. For instance, where the prior appropriation system allows the diversion of water for use on nonoverlying lands, the rights of groundwater users in designated basins are restricted to the use of groundwater on their overlying lands. Groundwater extracted in a designated basin "cannot be used to irrigate other lands without first receiving authorization from the Commission" (Fischer and Ray 1978, 52).

Also, where strict adherence to a priority system among pumpers might imply a "no injury" rule barring any pumping that lowers the water level for others, the rules for the designated groundwater basins allow some "mining" of groundwater under the Ground Water Commission's supervision. The commission's responsibility is to distinguish between reasonable and unreasonable impacts that one's actions may have on the water level for others (Fischer and Ray 1978, 52). Accordingly, in addition to giving the Ground Water Commission power to prohibit well drilling in designated basins, the 1965 Act also authorized it to regulate well spacing and to impose pumping limitations.

Where pumpers would prefer local control to state regulation of the basin, they may form a groundwater-management district, governed by a board of directors composed of basin residents and funded by property taxes within the district. Such a board assumes essentially the same powers and functions within the basin as the Ground Water Commission would have exercised. The districts have broad authority to regulate groundwater usage, but must consult with and answer to the commission pursuant to permits issued by the commission. Currently, the Ground Water Commission has established eight designated basins, and 13 groundwater-management districts within those basins (Colorado Ground Water Commission 2001).

Another set of rules applies to wells pulling groundwater that is not tributary to a natural surface stream, that is, nontributary groundwater. "These are waters held in natural basins underground from which they cannot escape" (Fischer and Ray 1978, 48). In nontributary groundwater areas outside of designated basins, wells and water uses are administered by the state engineer's office. The state engineer reviews new permit applications to make a judgment on the following considerations: (1) the extent of the supply in the basin, (2) the effect of existing wells on that supply, and (3) the goal of making the existing water supply last for 100 years. Permits are granted only for groundwater use on overlying lands within the basin (Fischer and Ray 1978, 57).

Water in Colorado is governed principally at the local level. The most important sources—native surface water flows and tributary groundwater—are governed by the prior appropriation system, which is administered at the local level through water appropriators, commissioners, courts, and division engineers. Transmountain diversions are controlled by the municipalities or conservancy districts that undertook them. Even groundwater in designated basins, which comes within the jurisdiction of the Colorado Ground Water Commission, can be managed by a local district if the residents prefer. Water governance and administration in Colorado is highly decentralized, with water users playing crucial roles in devising, overseeing, and enforcing water rights, rules, and regulations.

The Emergence of Conjunctive Management

Given the numerous physical and socioeconomic similarities among Colorado, Arizona, and California, it would be reasonable to suppose that conjunctive management in Colorado is similar to conjunctive management in Arizona and California. Water-scarce municipalities and irrigation districts are located in the driest section of the state, which also happens to be supplied with groundwater basins, surface storage reservoirs, and canals.

Although the eastern portion of Colorado contains the necessary components for the long-term storage of surplus surface water in groundwater basins, that type of conjunctive water management is not practiced. Instead, conjunctive management is used to maintain flows of surface streams. Maintaining surface streams protects senior surface water-rights holders and allows Colorado to meet the terms of interstate river compacts.

Development in eastern Colorado from the mid-1800s to the mid-1900s was accomplished almost entirely through the development of native surface water, supplemented by transmountain diversions. Since then, farmers in the region have embraced groundwater. Groundwater is relatively inexpensive and dependable compared with surface water. Use of groundwater allowed additional land to be irrigated at a lower cost rather than developing additional surface water supplies. Also, unlike surface water resources, the supply of groundwater is not limited during droughts. Finally, watering crops with sprinkler systems fed by wells is less labor intensive and more convenient than relying on irrigation tubes set by hand.

Many Colorado farmers came to appreciate the advantages of groundwater when drought gripped the state during the 1950s. Consequently, groundwater development continued even after the drought had subsided, with 1965 a peak year in the number of wells drilled (Vranesh 1987, *258*). Illustrating the rapid growth of the popularity of groundwater, the number of irrigation wells in the Arkansas basin skyrocketed from an estimated 40 in 1940 to 1,477 by 1972 (MacDonnell 1988, *582*).

Many of these wells pulled groundwater that was tributary to surface water streams. The effect of well pumping on surface water flows became apparent to many by the 1960s. Senior rights holders began demanding the shutdown of wells to protect their water rights, and at first glance they seemed to have the law on their side. According to the prior appropriation doctrine, when a senior appropriator calls for sufficient water to remain in the river to satisfy his water right, the appropriations of junior rights holders should cease until the senior appropriator's rights are satisfied. As the most junior appropriators, well owners should be shut down under these circumstances.

Three issues prevented such a direct solution. First, the Colorado constitution, legislature, and Supreme Court advocated the development and use of the waters of the state to the greatest extent possible for the benefit of the citizens of the state. Foreclosing the timely use of tributary groundwater violated such intentions. Second, the state engineer stated that he believed that he did not have the authority to regulate wells. Third, the concept of the futile call made it difficult, in practice, to shut down well pumping. A futile call occurs when a senior appropriator's rights would not be satisfied even if appropriations junior to it were shut down. In such a case, junior appropriators are allowed to continue to divert water. Shutting down wells to satisfy senior surface water calls is often futile because of a time lag between groundwater pumping and surface water flows. In most cases, shutting down wells will not have an appreciable effect on surface water flows for weeks or months. In many cases, a senior appropriator who made a call resulting in the shutdown of pumping would not realize any water for his crops until the irrigation season ended.

In 1965, the Colorado Legislature passed the Groundwater Management Act, which established a framework for governing groundwater. This Act created the Colorado Ground Water Commission (discussed earlier) to regulate pumping in designated basins, and directed the state engineer to address the conflict between tributary groundwater and surface water by devising rules and regulations (Radosevich et al. 1976, *138*).

In the summer of 1966, the Engineer exercised his new authority and ordered several wells in the Arkansas River basin shut down to satisfy senior rights holders with appropriations dating to 1887 (Radosevich et al. 1976, *139*). A lawsuit against the state engineer ensued and was eventually brought before the Colorado Supreme Court. The Court found that the Engineer had inappropriately exercised his powers and directed the Engineer to follow three steps in devising rules and regulations: (1) the rules and regulations must be based on a plan and a set of procedures; (2) the regulations must promise reasonable lessening of injury to senior rights; and (3) well owners should be allowed continued use of their wells if at all possible [Fellhauer v. People, 167 Colo. 320, 447 P.2d 986 (1969)].

The Colorado legislature acted to assist the state engineer in resolving the conflict between groundwater and surface water users by passing the 1969 Water Rights Determination Act. The 1969 Act clarified the priority system by creating a public ordering of all rights, provided incentives to well owners to adjudicate their tributary groundwater rights, and provided a mechanism by which well owners could avoid being shut down.

In 1969, no accurate and comprehensive listing of water rights in each of the water divisions existed. The division engineers were assigned the responsibility of tabulating existing water rights within each division. Persons with previously adjudicated rights had to come forward by 1974 and present evidence of their water-rights decree to preserve their priority. The division engineers now produce a tabulation of water rights for each division every two years.

As mentioned earlier, the Act designated a special period through June 30, 1972, during which a user of tributary groundwater could come forward and have his or her rights adjudicated with a priority date that would be the same as the true appropriation date. Since 1972, tributary groundwater rights have been determined and dated in the same manner as surface water rights. Colorado stands apart, not only from Arizona and California but also from most western states in this respect, having integrated surface and tributary groundwater rights into a single priority system.

In 1969, most wells were vulnerable to shutdown because they represented junior appropriations. Instead of shutting down wells, which since 1950 had rapidly become a major source of irrigation water, the 1969 Act provided for augmentation plans. Augmentation "provides a highly flexible tool enabling new uses of water without strict regard for the priority system, so long as existing rights are not injuriously affected" (MacDonnell 1988, 589). In other words, junior appropriators, whether of surface water or of tributary groundwater, can protect their diversions from "calls" by senior appropriators by augmenting stream flow.

A plan of augmentation for a well, or series of wells, involves first determining the depletions of stream flows, or injury to the river, caused by out-of-priority well pumping. Second, a source of water is identified that will be made available to the river at the time and place of injury. Augmentation plans ordinarily must be approved and decreed by water courts; temporary plans of augmentation, renewed on an annual basis, are administered by the state engineer's office (MacDonnell 1988). However, as discussed below, the authority of the state engineer to administer temporary plans of augmentation, also known as temporary substitute supply plans, very recently has been limited by court decisions and legislative action.

Conjunctive Management in the South Platte River Basin

Following several years of conflict among water appropriators in the South Platte basin, an agreed on set of rules regulating wells was accepted by the Division 1 Water Court on March 15, 1974 (Radosevich et al. 1976, *148*). The rules defined a timetable for phasing out well pumping, but allowed wells covered by an approved plan of augmentation to continue to operate. Key to incorporating tributary groundwater into the prior appropriation system were augmentation plans that allow out-of-priority depletions.

In the South Platte basin, which is Division 1 in Colorado's water administration system, some irrigation districts or companies have chosen to obtain decreed plans of augmentation or recharge to cover their members' wells. In our sample of water districts within Division 1, District 1 included six irrigation districts that had chosen to obtain decrees. Two districts chose to obtain decreed plans of augmentation—the Bijou Irrigation Company and the Fort Morgan Irrigation Company and District. The other four districts obtained decreed plans of recharge.

All told, these six irrigation organizations in District 1 of Division 1 own or lease and operate 39 separate recharge or augmentation sites that are used to cover the out-of-priority pumping of approximately 600 wells. Between 1980, when augmentation and recharge projects began coming on line, and 1997, the six organizations diverted 409,212 acre-feet of water into the various recharge sites.

Decreed Augmentation Plans

Decreed plans of augmentation include lists of wells to be covered, lists of augmentation structures to be used for recharging water to the aquifer and the river, the methods for measuring well depletions and augmentation accretions, and decreed rights of augmentation water. The operation, administration, and monitoring of the augmentation plan is shared among the irrigation districts, the District 1 water commissioner, the Division 1 engineer's office, and the Northern Colorado Water Conservancy District.

The irrigation districts with decreed plans of augmentation have incorporated augmentation activities within their irrigation infrastructure. For instance, the Fort Morgan Irrigation Company and District's augmentation structures consist of the Fort Morgan canal, several stretches of Badger Creek adjoining the canal, and several "prairie potholes" and ponds located adjacent to or at the end of the canal. During the nonirrigation season, October–March, the Fort Morgan district diverts water from the South Platte River under its augmentation decree when the decree is in priority, that is, when there is sufficient water in the river to satisfy a water right with a decreed date of May 19, 1972. The water is run in the augmentation structures, seeps underground, slowly flows back to the river, and enhances the stream flow of the river around the time of peak summer demand. This is precisely the time of year when pumping would be out-of-priority. Thus, the water from the augmentation structures replaces the water taken from the river by district members' wells. The wells can operate, even in the summer when they are out-of-priority, because of the replacement water that has been provided to the South Platte River by the Fort Morgan district's augmentation structures.

Decreed Recharge Plans

Decreed plans of recharge fall short of decreed plans of augmentation in that recharge plans do not fully cover out-of-priority pumping depletions. Recharge plans provide decreed amounts of water to be used for recharge, but stipulate that recharge credits must be used to cover out-of-priority depletions of members' wells—they cannot be leased or sold for any other purpose. Also, the organization holding the decree agrees to diligently enhance its recharge efforts so that eventually all well pumping will be fully covered. Once that occurs, the organization must seek a decreed plan of augmentation.

Decreed plans of recharge are common among the irrigation companies and districts of the South Platte Division. As mentioned previously, four of the six irrigation organizations in District 1 with decrees hold decreed plans of recharge. The organizations lease their recharge credits to Groundwater Appropriators of the South Platte (GASP), a well-owners' association of which they are members.

The recharge projects are similar to the augmentation projects operated by the Bijou and Fort Morgan irrigation companies. The organizations' irrigation ditches are often relied on as recharge sites. Irrigation ditches are essentially "free" infrastructure for recharge use: the companies are not required to make capital improvements to them to use them for recharge. During the irrigation off-season, generally October through mid-March, the companies run water through the ditches. To gain recharge credits, no one along the ditch may irrigate. Instead the water in the ditches is allowed to percolate through the bottom of the ditch into the soil and work its way gradually toward the river.

Other sites used for recharge plans are almost as simple. They consist of creek beds with check dams, leaky reservoirs, and naturally occurring depressions preferably next to or at the end of the irrigation ditch. At most, a company might expand the capacity of a naturally occurring depression by building earthen berms.

The ditches and ponds are at varying distances from the South Platte River. The farther a recharge pond is from the river, the longer it takes for the water from the pond to migrate back to the river. To return water to the river at various critical times, water is distributed among recharge sites of varying distances. Recharged water moves toward the river quite slowly; water from a company's sites will continue to flow into the river for three to four years after the company ceases recharge operations.

Temporary Augmentation Plans

Water appropriators and the state engineer recognized the need for a mechanism that would allow well owners to continue to pump even as they sought decreed plans of augmentation and recharge, or even if they did not. In 1972, with the encouragement of the state engineer, a group of well owners formed the GASP as a nonprofit organization to develop a portfolio of water that could be used to cover stream depletions caused by members' out-of-priority well pumping. The organization agreed to provide the state engineer with a list of its members and their wells, an estimate of the amount of water to be pumped in the coming irrigation season, and the amount of water pumped in the previous irrigation season, and to make available to the state engineer an amount of water to replace out-of-priority depletions and offset injuries to senior rights (MacDonnell 1988, *591*). The state engineer accepted the offer. These temporary augmentation plans, or substitute supply plans, must be approved by the state engineer on an annual basis. The state engineer has approved a substitute supply plan for GASP for each of the last 30 years.

Using revenues raised through membership fees, GASP leases shares of water from ditch companies and reservoirs, and recharge credits from irrigation districts such as those in District 1. The organization combines these sources to cover out-of-priority well pumping over a 200-mile reach of the South Platte River. GASP has never entered water court to obtain water decrees, or shown any intention of formalizing its water portfolio. GASP operates instead on a year-to-year basis with the approval of the state engineer.[4]

"The GASP approach has been characterized as 'call management'"(Mac-Donnell 1988, *592*). The GASP water portfolio is sized and located so as to "minimize the call on the lower portion of the South Platte River" (MacDonnell 1988, *612*). Technically, call management violates the prior appropriation doctrine, because it does not fully replace out-of-priority stream depletions. Call management simply quiets the protests of the senior appropriators most likely to complain about well pumping. What the state engineer and GASP managed to reclaim was the status quo on the South Platte River prior to the widespread use of wells. Those senior appropriators whose water rights historically were consistently satisfied, especially in the middle of summer, continue to have their rights protected from well pumping.

Substitute supply plans and the state engineer's willingness to approve them year after year without requiring well owners or associations to develop court-decreed plans of augmentation were never viewed favorably by senior surface water rights holders and by districts and ditch companies that invested the time and resources to obtain court-decreed plans of augmentation. Senior water-rights holders feared that if a severe drought struck, the state engineer would be reluctant to shut down some or all of the thousands of wells covered by substitute supply plans in order to protect senior water rights. Districts and ditches with decreed augmentation plans believed that all well owners should be subject to the same rules. Even though they were dissatisfied they did not challenge temporary substitute supply plans in court. As it turned out, they did not have to.

A court case, involving a water conflict in the Arkansas River Basin, that on its face did not challenge substitute supply plans, coincided with one of the most severe droughts on record to bring an end to the practice of the state engineer approving substitute supply plans in the South Platte River basin. At the end of 2001, the Colorado Supreme Court ruled that a party to a water conflict did not have standing to bring suit because it did not hold an adjudicated water right, only a temporary substitute supply plan approved by the state engineer [Empire Lodge Homeowners' Ass'n v. Moyer 39 P.3d 1139 (Colo. 2001)]. Further, the Supreme Court ruled that the state engineer did not have the authority to approve substitute supply plans.

To save the 2002 irrigation season in the South Platte basin, the state legislature passed a law allowing the state engineer to approve existing temporary substitute supply plans for one year, and delineated very limited circumstances under which the state engineer could approve new plans. For instance, the state engineer could approve a substitute supply plan only if the applicant had also filed for a decreed plan of augmentation in a water court (H.B. 02-1414; Colo. Rev. Stat. §37-92-308). In May 2002, the state engineer promulgated a new set of rules regulating groundwater pumping in the South Platte River that would allow the engineer to approve replacement plans "whereby out-of-priority depletions could be replaced under an administratively approved plan subject to public notice and an opportunity for the public to comment on the adequacy of the plan" (Paddock and Hammond 2003, 6).

The flurry of legislative and administrative activity occurred during the worst water year on record (Byers 2002). Senior water-rights holders challenged the proposed rules in water court. At the end of 2002, the water court decided in their favor, ruling that the state engineer lacked the authority to approve out-of-priority depletions under replacement plans (in re the Proposed Amended Rules and Regulations Governing the Diversion and use of Tributary Ground Water in the S. Platte River Basin, No. 02-CW-108 D.C. Water Div. No.1. Colo. Dec. 23, 2002). The Colorado Supreme Court affirmed the decision of the water court [Simpson v. Bijou Irrigation Co., 69 P.3d 50(Colo.2003)]. On the same day that the Supreme Court issued its decision, April 30, 2003, the governor of Colorado signed into law a bill that authorizes the state engineer to issue annual approvals of substitute supply plans for three more years. After that, substitute supply plans may be approved only if an augmentation plan has been filed with a water court. During the three-year grace period, the state engineer may approve plans only after he or she holds hearings on the plans and provides extensive written reports detailing the factual basis for the plan and identifying terms and conditions that will prevent injury to senior water rights (Senate Bill 03-073).

The conflict among water users in the South Platte is intense. Well owners claim that replacement water will be scarce and expensive, forcing them to pump less to reduce the effect on the river, negatively impacting the $130 million agriculture economy (Smith 2003). Senior water-rights holders counter that during times of shortage, such as a drought, they are forced to reduce their water use and that well owners should be subject to the same constraints (Smith 2002).

Instream Flows and the South Platte River Compact

Satisfaction of senior surface water rights is not the only impetus for conjunctive management in the South Platte River basin. Another is the need for increasing instream flows to protect endangered species. The U.S. Fish and Wildlife Service (USFWS) has listed the whooping crane, the least tern, the piping plover, and the pallid sturgeon as endangered. All four species use the central Platte River region located in Nebraska. Wyoming, through which the North Platte River flows; Colorado; Nebraska; and the U.S. Department of the Interior have a cooperative agreement to develop and implement a recovery program for the four species (Cooperative Agreement 1997). One aspect of the recovery program focuses on water flows. The USFWS has developed recommended flows needed at different times of the year by endangered species. Initially, the three states are attempting to change the timing of flows for approximately 70,000 acre-feet of water, capturing water when flows exceed USFWS recommendations and releasing them when flows fall short. Colorado has committed to making available 10,000 acre-feet of water between April and September of each year by adjusting the timing of water flows (Aiken 1999).

Colorado intends to meet its commitment through a recharge program located at the Tamarack Ranch State Wildlife Area, 40 miles from the Nebraska border. The recharge program is being developed and administered through a coalition of Front Range cities, GASP, the Northern Colorado Water Conservancy District, and the Colorado Department of Fish and Wildlife. The Tamarack recharge program consists of a series of wells and recharge ponds. The wells, located very near to the South Platte River, are pumped in the early winter and spring when there is surplus water in the river. The water is placed in recharge ponds at differing distances from the river. The water percolates into the tributary groundwater aquifer, and returns to the river between April and September when the supplemental stream flows are needed to satisfy Colorado's commitment to the species recovery program (Cooperative Agreement 1997).

Conjunctive Management in the Arkansas River Basin

Augmentation in the Arkansas River Basin (Division 2) is conducted for similar purposes as in the South Platte basin (Division 1), but in an entirely different manner. Division 2 well owners have incorporated tributary groundwater into the prior appropriation system differently, and the differences stem from a combination of physical and institutional circumstances.

Arkansas basin appropriators did not at first respond to the tributary groundwater crisis by adopting augmentation. Instead, a set of rules regulating well pumping was adopted in 1973. The rules limited pumping to three days per week—Monday, Tuesday, and Wednesday. In 1974, the division engineer attempted to adopt the South Platte basin approach of phasing out well pumping unless wells were part of an approved augmentation plan. The well owners challenged this rule in court, and the Colorado Supreme Court sided with them, concluding that the engineer had not demonstrated that such measures would make additional water available to senior appropriators (in re Arkansas River, 195 Colo. 557, 581 P.2d 293, 1978).

The three-day-a-week rule did not limit pumping, in part because it was not enforced. Other mechanisms, however, did ensure an upper bound on pumping. Under the 1969 Water Rights and Administration Act, well owners adjudicated their wells. Each decree defined the volume of water that could be put to beneficial use, which placed limits on pumping. Furthermore, the division engineer ceased issuing permits for new irrigation wells. This was the status quo until the mid-1980s, when Kansas filed suit in the United States Supreme Court against Colorado, claiming that Colorado had violated the Arkansas River Compact by depleting usable river flows into Kansas by the operation of two reservoirs in Colorado and by increased groundwater pumping. The Court-appointed Special Master found that Kansas failed to demonstrate the impact of the two reservoirs, but sided with Kansas in finding that increased pumping had depleted Arkansas River flows.[5] The Supreme Court accepted the Special Master's findings. (Kansas v. Colorado, 514 U.S. 673, 1995).

The State of Colorado acted quickly to bring Arkansas River wells within the prior appropriation system to minimize the penalties the state would have to pay Kansas. Similar to what had transpired in the South Platte basin two decades earlier, the state and division engineers, the state attorney general, and the well-owners' associations, within the context of the water court, devised a set of rules to regulate pumping.

The rules created replacement plans, which are a cross between decreed plans of augmentation and temporary plans of augmentation. Replacement plans are similar to decreed plans in that they fully replace each out-of-priority depletion at the time and point of injury. But, like temporary plans of augmentation, replacement plans are not adjudicated; they are approved by the division engineer each year. The determination of out-of-priority depletions is based on assumptions of irrigation use and efficiency. Where flood irrigation is used, the amount of pumped groundwater that must be replaced varies from 30 percent, if the groundwater is used to supplement surface water supplies, to 50 percent if groundwater is the sole source of irrigation water. Where sprinkler irrigation is used, 75 percent of pumped groundwater must be replaced (Colorado State Engineer's Office 1996).

Each year, well-owners' associations provide the division engineer with a list, by river reach, of wells and the amount of water each well is expected to pump, and the water that the association will make available to the engineer to replace stream depletions from out-of-priority pumping. The engineer's office collects monthly data on well pumping, stream depletions, and stream replacements. Each month, the division engineer, the associations, and a representative from the State of Kansas review the accounts to ensure that stream depletions from out-of-priority pumping have been covered.

The replacement plans developed by the well-owners' associations are in-lieu recharge programs. Instead of recharging water into the aquifer directly, as do some of their counterparts in the South Platte basin, the Division 2 well-owners' associations purchase or lease rights to surface water. The surface water is released to the stream over the course of the irrigation season to replace the water depleted by out-of-priority well pumping. The major source of replacement water comes from the Arkansas-Frying Pan Project, owned and operated by the Southeastern Colorado Water Conservancy District. Between 1975 and 2001, the conservancy district on average delivered 63,800 acre-feet of water each year to farmers for irrigation purposes. The return flows from irrigation using conservancy water are owned by the conservancy district.[6] The district leases return flows to the well associations to use as replacement water. Almost 50 percent of all replacement water is comprised of return flows (Kepler 2003). The remaining replacement water comes primarily from surface projects developed by the cities of Pueblo and Colorado Springs, and "first use," as opposed to return flow, water from the Arkansas-Frying Pan project. Well-owners' associations have also purchased shares of water of mutual ditch companies (Colorado Water Protective and Development Association Augmentation Plan 1998).

Well-owners' associations along the Arkansas River have chosen to develop in-lieu recharge programs partly because of their physical circumstances. First, the tributary aquifer of the Arkansas River is narrower than that of the South Platte, and water tables are higher. Thus there are fewer opportunities to recharge into the tributary aquifer, and less assurance that recharged water will return to the river. Second, the Arkansas River is under a "call" year around. Only rarely would a very junior augmentation decree be in priority so that water could be drawn from the river and placed in recharge ponds. Third, cities located upstream of the well owners have developed surface storage systems whose volume currently exceeds their water needs. Cities have surplus surface water to lease, and probably will have for the next 50 years, except in times of drought.

Three well-owners' associations were captured in our sample from Division 2. They are the largest well associations in the basin, covering approximately 95 percent of all wells pumping tributary groundwater. The wells range from Pueblo in the west to the Kansas border in the east. Two of the associations—the Arkansas Groundwater Users Association (AGUA) and the Colorado Water Protection and Development Association (CWPDA)—cover all wells between Pueblo and the John Martin Reservoir, just west of Lamar. These two associations cover approximately 1,800 wells. From 1996, when they first replaced water, through 1998, the associations made available an average of 44,000 acre-feet of replacement water per year. The Lower Arkansas Water Management Association (LAWMA) covers 640 wells between the John Martin Reservoir and the Kansas border. Most of the wells covered by LAMA lie outside of the boundaries of the Southeastern Colorado Water Conservancy District; thus the association does not have access to Arkansas-Frying Pan Project water. Instead, with loan assistance from the state it purchased shares of water from mutual ditch companies. The ditch shares are equal to approximately 14,000 acre-feet of water per year (Higbee 2000, personal communication).

The well owners of the Arkansas River basin have been immune to the conflict unfolding in the South Platte River basin, however, they have not been immune to drought. When the Colorado legislature allowed the state engineer to approve temporary substitute supply plans for three more years, in the wake of the 2003 Supreme Court ruling striking down the rules governing groundwater pumping in the South Platte, discussed in the preceding, the legislature also ratified the rules providing for replacement plans in the Arkansas River (Paddock 2003). Although the Arkansas replacement plans are approved annually by the state engineer and are not adjudicated in water court, thus apparently violating the decision of the Supreme Court, the replacement plans are an integral part of the settlement of the lawsuit brought by Kansas concerning the Arkansas River Compact. Thus, the legislature in protecting the replacement plans, acted to protect the settlement.

The severe drought that has plagued Colorado for the past several years, however, demonstrates the fragility of the replacement plans. Replacement plans depend on a source, or sources, of surplus surface water. If such sources are not available, then pumping must be restricted to a level that does not affect river flows. With the current drought, the well owners in the Arkansas River Valley do not have access to their usual supplies of replacement water. Replacement water for irrigation wells in 2003 was only a fraction of that available in prior years. Consequently, well pumping was strictly limited, primarily for essential municipal, domestic, and livestock uses (Kepler 2003).

Conclusion

Over the course of just over three decades, from 1970 through 2003, eastern Colorado water users struggled to incorporate tributary groundwater within the prior appropriation system. Tributary groundwater users in the South Platte Basin use direct recharge basins and irrigation ditches and canals for their augmentation and recharge plans. Water from the South Platte and its tributaries diverted into these structures during the wetter season seeps into the tributary groundwater basin and migrates back to the river, much of it reaching the river during the summer. Well-owners' associations such as GASP have based their substitute supply plans on leasing surplus surface water and/or recharge and augmentation credits to cover their members' out-of-priority pumping. With these methods, the tributary groundwater crisis that emerged in the early 1970s in the South Platte basin appeared to have been largely resolved by the end of the same decade. Although some questioned the adequacy of substitute supply plans, after 1990 most attention turned to developing recharge projects near the Nebraska border to meet Colorado's obligations to the Platte River system and to protect endangered species in the Central Platte region. However, with the reappearance of drought in 2000, the truce between senior, mostly surface water, rights holders and junior well owners crumbled. Within three years, all out-of-priority depletions to the South Platte River must be covered by augmentation plans decreed and administered by the water court, placing all water users of native surface flows and tributary groundwater on equal footing within the prior appropriation doctrine.

In the Arkansas Basin, the tributary groundwater crisis erupted in the 1990s. The U.S. Supreme Court's decision forced tributary groundwater users to rapidly bring their well pumping within the prior appropriation system. Well-owners' associations leased and/or purchased surface water to cover their out-of-priority pumping. The associations continue to work to firm up their sources of replacement water, as the availability of sources currently leased from upstream cities will be restricted by the cities' growth and certainly by droughts.

Colorado's practice of conjunctive management differs substantially from Arizona's or California's. The differences result from the interactive effects of Colorado's reliance on surface water for irrigation and municipal supplies, and the state's commitment to the prior appropriation system for allocating water among users. The combination of those historical and institutional factors caused groundwater to be developed comparatively late in Colorado, and have led the state and its water users to look on groundwater primarily as a source of supplemental supply. Thus, the institutional arrangements governing and implementing conjunctive management in Colorado largely place groundwater, and the junior appropriators who rely upon it, in the service of maintaining surface stream flows.[7]

Conjunctive management in Colorado allows groundwater users—primarily irrigators and some newer municipalities and subdivisions—to access and use tributary groundwater that otherwise would not be available if the prior appropriation doctrine were strictly applied. Virtually all well owners are junior to most surface water users, and thus in the overappropriated South Platte and Arkansas river basins their water rights would rarely be in priority. Conjunctive management was a way of keeping the prior appropriation system and allowing tributary groundwater use to continue. Rather than modify or replace the prior appropriation system, Coloradoans searched for and invented ways to accommodate at least some of the more recent groundwater use while protecting stream flows for senior surface water users.

Conjunctive management now also allows Colorado to meet water obligations to downstream states. Supplementing the flows of the South Platte and Arkansas rivers has enabled Colorado to satisfy its river compact obligations to Kansas and Nebraska, and also to fulfill its agreement with the federal government and the states of Nebraska and Wyoming to restore endangered species along the Platte River.

For the immediate future, Colorado appears to have once again resolved the conflict between surface water users and groundwater users. However, Colorado's commitment to the prior appropriation doctrine is likely to be challenged in the future as the state's population continues to grow unabated and as well owners find it increasingly difficult and costly to find surplus surface water to replace out-of-priority well pumping. Continued growth in demand for water combined with periodic droughts will increase pressure to use the millions of acre-feet of groundwater tributary to the South Platte and Arkansas Rivers that is currently unavailable for use because of the impact that such widespread pumping would have on stream flows. Tributary groundwater can be accessed only to a very limited degree as long as the prior appropriation doctrine is used to protect surface stream flows.

Part III

Institutions and Policy Change: Analysis and Recommendations

7
Tracing and Comparing Institutional Effects

It is not surprising that conjunctive management is practiced in Arizona, California, and Colorado. All three states have the necessary physical components of groundwater basins that can be combined with surface water impoundment and conveyance facilities. In all three states, growing populations and intensive agriculture have strained the existing and erratic water supplies that depend on annual precipitation and runoff, furthering the search for drought-resistant water management alternatives. Especially in an era when more big dams are widely understood to be as unlikely as they are undesirable, implementing conjunctive management seems to be an obvious step for these three states. What is not as obvious is why conjunctive management practices have followed such divergent paths from California to Arizona to Colorado. As each state's conjunctive management experience was detailed in Part 2, differences in the purposes, methods, organizational structures, and magnitude of conjunctive management emerged despite their relatively similar water resources and problems. Some key elements of conjunctive management in the three states, which we discuss throughout the chapter, are captured in Table 7-1.

Without question, differences among the states result partly from physical and historical distinctions, but to a greater degree they illustrate how institutions matter. The differences among the states are related to institutional arrangements such as water-rights laws and the assignment of responsibilities for governance and management of water supplies. These institutional factors also account for some aspects of conjunctive management practice where two states are similar but the third is distinct. Here we trace and compare how institutional arrangements in the three states account for their similarities and differences.

Table 7-1. Comparison of Conjunctive Management Institutions, Purposes, and Organizations

	California	Arizona	Colorado
Conjunctive management purposes			
Conjunctive management is used to enhance stream flows			X
Conjunctive management is used for seasonal storage	X	X	X
Conjunctive management is used for long-term storage	X	X	
Conjunctive management is used for recovery of groundwater overdraft	X	X	
Conjunctive management methods			
Groundwater users store water to recharge streams and offset pumping effects			X
In-lieu recharge—incentives offered to use surface water when plentiful, allowing groundwater storage to increase	X	X	
Direct recharge—surplus surface water sunk or injected underground for later capture	X	X	
Water-rights systems			
Integrated ground and surface water rights			X
State law quantifies groundwater rights		X	X
State defines rights to recharge and recover stored groundwater		X	X
State permits local-level institutions to define rights to groundwater	X		
Organizational arrangements			
State agencies approve locally organized conjunctive management projects		X	X
State agencies operate conjunctive management projects		X	
Private organizations and individuals participate in conjunctive management projects	Rarely	X	X
Substate governments, including municipalities, counties participate in conjunctive management	X	X	Rarely
Organizations providing conjunctive management coordinate around basin boundaries	X		

Institutional Arrangements and the Purposes of Conjunctive Management

Why do people undertake conjunctive management of surface and groundwater resources, and do institutional factors affect those choices? As described in Part 1, conjunctive management can serve a number of water resource management purposes. One might well expect that the purposes considered most important in a particular location would depend on physical conditions, and surely that is important. Our study of Arizona, California, and Colorado, however, reveals that the reasons people pursue conjunctive management are driven also, perhaps even more, by the institutional milieu within which they use and manage water supplies—water-rights laws, economic incentives and constraints of water projects, even interstate water allocations and obligations. Institutional arrangements not only explain how conjunctive management has been pursued in these three state, but institutions also help explain why conjunctive management has been pursued in the first place.

Tracing these effects of institutional arrangements on conjunctive management purposes helps us understand several ways in which Colorado is distinct from Arizona and California, as well as some ways in which Arizona and California are different from one another. In Arizona and California, the primary purpose of conjunctive management is the long-term underground storage of surplus surface water. The stored water is called on to buffer urban populations from drought, to meet anticipated future water demands, and to provide large-scale water project operators with additional prospects for water sales in wet years and supplemental supplies to recover in dry years. Conjunctive management in Arizona and California is also pursued to restore water levels in overdrafted aquifers, and an additional but localized purpose in California is to protect coastal aquifers from seawater intrusion.

In Colorado the primary purpose of conjunctive management is not long-term storage of surplus surface water, or recovery of overdrafted basins. Conjunctive management is undertaken to ensure adequate stream flows to protect surface water appropriators' rights, to allow groundwater pumping to continue to supply many of the newer communities and farms in the state, and to meet obligations to downstream states.

These differences of purpose are linked to institutional arrangements developed in response to water resource conditions and problems that emerged during each state's development. Reliance on groundwater occurred relatively early in Arizona and California—especially southern California—and signs of groundwater overdraft were apparent on both states by the 1930s. Southern California water-user associations, and the Arizona legislature, began to seek ways of restricting groundwater use in some of the most severely affected basins. Cutting back pumping was inconsistent, however, with dependence on groundwater to supply each state's rapidly growing population and economy. California and Arizona water-user organizations and government officials sought imported supplies to supplement groundwater sources. Financing those large-scale importation projects required steady revenues from water deliveries, so project operators were also interested in finding ways to smooth out variations in water demand and supply. Partly to achieve improved water-resource management and partly to meet the financial requirements of those large-scale water importation projects, the surface water delivered by those projects came to be managed conjunctively with groundwater resources in the two states.

The logic of conjunctive management was clear enough, but institutional arrangements in both California and Arizona impeded its large-scale development, leading water users to devise new institutional arrangements that would facilitate their efforts. At mid-twentieth century, water law as interpreted and modified by courts in both states did not recognize the hydrologic connections between ground and surface water. Percolating groundwater was governed by the beneficial use doctrine, while water flowing in surface or underground channels followed the prior appropriation doctrine. Neither state's water laws adequately recognized and protected rights to store and recover underground water.

The state of water law in California and Arizona created two principal obstacles to conjunctive management. One problem was the incommensurability of water entitlements and liabilities among surface and groundwater users, and the structure of incentives that resulted. Surface water users were largely unprotected from groundwater pumping that depleted the flows of hydrologically connected streams. Because by law, pumpers of percolating groundwater were using a different type of water, they were not liable for drying up surface streams.[1] On the other hand, groundwater users were largely unprotected from surface water users' impoundments and diversions, which removed water from the natural channels where it percolates underground and replenishes aquifers. Surface water sources of any note that were within reach of agricultural and urban areas were dammed and diverted until dry, and the overdrafting of groundwater continued apace. Furthermore, the difference between the appropriative rights system used for surface water and the beneficial use doctrine governing groundwater robbed surface water-rights holders of any incentive to participate in a conjunctive management arrangement—regardless of hydrologic conditions, why would they ever exchange surface flows in which they had quantified and protected rights for use of a relatively open-access groundwater source?

The other problem with Arizona and California law was the lack of assurance among water users, especially groundwater users, that water conserved and stored underground could be accounted for, preserved, and recovered. Without either quantified rights to groundwater use or rights to underground water storage, groundwater users had little incentive to restrain water use in the present and save it for the future, which is essential if the primary purpose of a conjunctive management program is long-term storage of water to reduce exposure to drought.

These problems generated by the water-rights systems in Arizona and California further shaped the purposes of conjunctive management in each state. In Arizona, these concerns combined with state officials' desire to secure federal support for the Central Arizona Project (CAP) so Arizona could use its allocation of Colorado River water. Securing CAP support meant resolving some of the state's other water supply dilemmas. As recounted in Chapter 5, state officials and representatives of the major water user groups changed state water law in 1980 to establish quantified groundwater rights and limitations, close the most heavily used basins to additional pumping, and set a goal of ending groundwater mining.

The new system of quantified and secured groundwater rights in Arizona added some assurance that groundwater left unused today would still be available tomorrow. In addition, however, the new law's restrictions on groundwater mining and its emphasis on recovering overdrafted basins contributed to the need to put the new surface supplies from the CAP to use, thus lending a new dimension to the purposes of conjunctive management in Arizona. The institutional changes that yielded the CAP but arrested groundwater mining also established the goals of conjunctive management in Arizona today—storing large quantities of surplus surface water, primarily project water, underground to make use of the state's Colorado River allocation, save water for future needs, and aid in the recovery of overdrafted basins. This connection between institutional arrangements and the purposes of conjunctive management in Arizona is summarized in the top half of Figure 7-1.

In California, the lack of assurance concerning stored water supplies also shaped the purposes of conjunctive management, but in a different direction. Groundwater users in California operated without quantified rights of withdrawal or rights to store and recapture, and furthermore in a setting where state officials' inability or unwillingness to choose among competing legal doctrines made it difficult for users to rely on *any* assurances about their rights or their ability to reap future benefits from present restraint. In many locations throughout the state, it was unclear to water users how their long-term future would be secured.

In this atmosphere of uncertainty, and facing critical overdraft conditions, water users in several California basins turned to conjunctive management. The promise of groundwater replenishment with supplemental water supplies, primarily project water, made more tractable the otherwise zero-sum nature of limiting groundwater use. A recharge program allowed water users within a basin to increase the size of the pie they were dividing. It also raised water users' confidence that basin conditions would not deteriorate further while they re-

Institutional factors at work in Arizona

1. Prior appropriation system for surface water supplies means surface water within the state is fully allocated to senior rights holders, leaving emphasis on groundwater and imported water to accommodate growth since 1950.

2. 1980 Groundwater Management Act identifies Active Management Areas (AMAs) and establishes substantial state-level policymaking role for them— setting safe-yield goals, quantifying and limiting pumping rights in AMAs, and requiring demonstration of assured water supplies for new development.

3. Financial viability of completed Central Arizona Project (CAP) and claims to Arizona's full share of Colorado River water necessitate maximizing sales and use of CAP water, even in anticipation of demand.

Contributing to
Purposes of conjunctive management in Arizona

1. Recovery of overdrafted basins in the AMAs and use of their storage capacity.

2. Allowing growth and urbanization to continue with assured water supplies, through banking, and in some instances transfers, of stored water.

3. Assuring that CAP water is purchased and stored for future in-state needs or interstate exchanges.

Which help explain
Institutional effects characteristic of Arizona conjunctive management

1. State laws and regulations facilitating groundwater storage and recovery, recognizing rights of water users to store and recapture underground water supplies.

2. State agencies that purchase and bank water underground in large-scale projects, including interstate arrangements with California and Nevada.

3. Cities and/or local water districts often undertaking banking projects on their own, owing to assured rights to store and recapture water.

But not basin-level groundwater management organizations.

Figure 7-1. Arizona Conjunctive Management Overview

strained their pumping, and held out the longer-term promise of recovered water levels, and in coastal basins the protection of water quality.

This helps to explain why most basins in California that have any groundwater management at all have some form of conjunctive management. The prevailing uncertainty in the absence of statewide rules governing groundwater use led users in each basin to try to find their own ways of solving overdraft problems. In California, groundwater management and conjunctive management are almost impossible to decouple: some control over a groundwater basin had to be developed to make a conjunctive management program sustainable, but introducing supplemental water supplies as part of a conjunctive management helped make it possible for groundwater users to agree on a basin management program. It is not too much to say that in many California basins, conjunctive management made groundwater management possible, and vice versa.

The influence of state water-rights law also helps explain why conjunctive management programs in California have been pursued first in urbanized and rapidly urbanizing groundwater basins. Agricultural groundwater users are generally pumping water for use on their own land, and those uses are protected fairly well under two of the competing groundwater use doctrines outlined in Chapter 4, specifically beneficial use and correlative rights. Those doctrines still leave uncertainty about the quantities of use rights and about storage and recapture, but the basic claim of a right to the use of the water is clearer. In urban areas, more of the groundwater doctrines come into play—not only beneficial use and/or correlative rights but appropriation by water suppliers for use on others' land, for example, private water companies and municipal utilities; prescriptive rights; and pueblo rights. As uncertainty about rights contributed to the impetus to establish some control over a basin, that impetus was greater in urbanized and rapidly urbanizing locations.

The desire of California groundwater users to arrest overdraft coincided with the interest of project water purveyors to find ways of making their water supplies attractive and financially viable even in periods of surplus. The project operators' interests and incentives were also guided by institutional arrangements regarding water rights and the repayment of project obligations. For example, the City of Los Angeles in its diversions from the Owens Valley and Mono Lake, and the Metropolitan Water District of Southern California in its diversions of Colorado River water, needed to establish relatively consistent annual surface water diversions to perfect and protect their entitlements to the water they were importing. Metropolitan has also adopted a system of "take-or-pay" contracts with its member agencies, requiring Metropolitan to deliver fixed annual amounts and requiring the members to pay for the water whether they need it or not. Similarly, contracts between the California Department of Water Resources and local agencies for State Water Project deliveries involve consistent annual payments to satisfy revenue bond obligations regardless of water-demand conditions. Thus the agencies involved in these water importation and diversion projects had institutionalized incentives to find ways to deliver and sell project water that did not depend entirely on water demands that fluctuate with weather conditions. In wet years when demands for the project water itself were low, using project water for groundwater replenishment satisfied the need of project operators to maintain consistent sales. Even at a discount, selling water to basin users who would sink it underground as part of a replenishment program was preferable to no sale. Project water is not used in every conjunctive management program in California, but the marriage of interests between project water purveyors and groundwater users has aided the development of some of the largest projects in the state.

Institutional arrangements have thus shaped the purposes for which conjunctive management has been pursued in California—not just how it is done, but why it is done at all. In certain respects, conjunctive management in the state has been a means of overcoming the challenges to effective groundwater management posed by the uncertainties and contradictions of the state's water-rights rules. Water users in the California basins have undertaken conjunctive management to resolve those uncertainties, at least within their own basin; overcome past overdrafting; provide a buffer against variations in surface supplies; and in some cases also ensure a more nearly steady demand for project water supplies. These connections between institutions and the purposes of conjunctive management in California are presented in the top half of Figure 7-2.

As described in Chapter 6, developments in Colorado were quite different from those in Arizona and California, and those differences explain why the purposes of conjunctive management there are distinct. Here too the rules governing surface water use—prior appropriation—differed originally from those governing groundwater use— beneficial use—but the sequence of water resource development in Colorado created a different dynamic in response to that separation. The use of groundwater for irrigation and municipal purposes occurred relatively late in the development of the state—around the mid-twentieth century—and by this time the streams and rivers in eastern Colorado were fully appropriated and major surface water projects had been completed.

Groundwater use emerged in an institutional environment designed by surface water appropriators to support and protect their water rights. Although several surface water users also had begun pumping groundwater and did not want to see pumping shut down completely, the century-old institutional arrangements that supported senior surface water rights holders were not going to be overthrown to accommodate groundwater users. Instead, beginning in the late 1960s, rights in tributary groundwater were cobbled into the existing institutional arrangements for surface water. That meant allowing groundwater pumping only if it did not harm senior surface water-rights holders. Conjunctive management in the form of augmentation, recharge, and replacement plans allowed for continued use and development of groundwater while protecting senior water-rights holders.

More recently, conjunctive management in eastern Colorado has also allowed the state to meet its interstate water compacts and commitments to protecting endangered species by boosting stream flows at the state borders. Thus, institutional arrangements such as the water-rights rules protecting senior surface water users, the integration of tributary groundwater rights into that system in a junior, inferior status, and the need to satisfy contractual and/or reg-

Institutional factors at work in California

1. State laws and court precedents allocate water rights by multiple theories, generating uncertainty about allocations during shortages.

2. State rules do not establish groundwater storage and recapture rights, nor do state agencies promote or govern conjunctive management.

3. "Home rule" establishes and empowers local government units, including special districts, as requested.

4. Financial rules, combined with use of contracts for purchase and delivery of project water, require imported water projects to deliver water, even in anticipation of demands.

Contributing to
Purposes of conjunctive management in California

1. Recovery of overdrafted basins and use of their storage capacity.

2. Allowing large-scale water project operators to store and transfer water to satisfy contractual obligations during dry periods.

3. Reduction of uncertainties over allocation among basin users in times of shortage by creating a cushion of stored water.

4. Allowing growth to continue where desired.

But not much emphasis on assuring stream flows for senior rights holders.

Which help explain
Institutional effects characteristic of California conjunctive management

1. Creation of basin-scale organizations to represent groundwater users and coordinate conjunctive management projects.

2. Large-scale water project operators involved primarily as sellers of replenishment water rather than project managers.

3. Lengthy and expensive processes, often involving litigation, leading to development of conjunctive management projects, because of the number of assurances that have to be secured.

4. Substantial variation in organizational structure and operation of conjunctive management projects from basin to basin.

Figure 7-2. California Conjunctive Management Overview

ulatory obligations to downstream interests account for the principal purpose of conjunctive management in Colorado, which is not the underground storage of water but the use of groundwater to supplement surface stream flows.

This connection between institutional factors and conjunctive management objectives in Colorado is summarized in the top half of Figure 7-3. Together, Figures 7-1, 7-2, and 7-3 highlight the connections between institutional arrangements in the three states and the purposes for which conjunctive management has been undertaken.

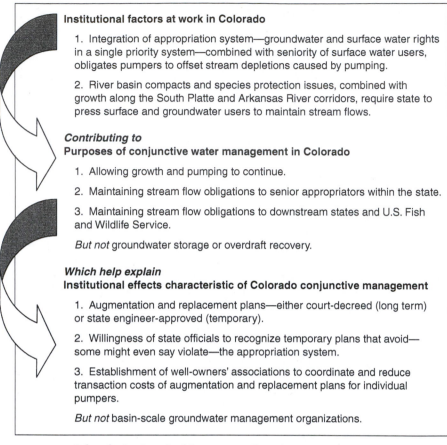

Institutional factors at work in Colorado

    1. Integration of appropriation system—groundwater and surface water rights in a single priority system—combined with seniority of surface water users, obligates pumpers to offset stream depletions caused by pumping.

    2. River basin compacts and species protection issues, combined with growth along the South Platte and Arkansas River corridors, require state to press surface and groundwater users to maintain stream flows.

Contributing to
Purposes of conjunctive water management in Colorado

    1. Allowing growth and pumping to continue.

    2. Maintaining stream flow obligations to senior appropriators within the state.

    3. Maintaining stream flow obligations to downstream states and U.S. Fish and Wildlife Service.

But not groundwater storage or overdraft recovery.

Which help explain
Institutional effects characteristic of Colorado conjunctive management

    1. Augmentation and replacement plans—either court-decreed (long term) or state engineer-approved (temporary).

    2. Willingness of state officials to recognize temporary plans that avoid—some might even say violate—the appropriation system.

    3. Establishment of well-owners' associations to coordinate and reduce transaction costs of augmentation and replacement plans for individual pumpers.

But not basin-scale groundwater management organizations.

Figure 7-3. Colorado Conjunctive Management Overview

Institutional Arrangements and the Practice of Conjunctive Management

In addition to their influence on the purposes of conjunctive management, institutional factors have shaped the means by which conjunctive management is performed, that is, the types of projects found in the three states. Comparison along this dimension highlights additional differences between Arizona and California, as well as Colorado's distinctiveness from both. The purposes of conjunctive management and the means by which it is conducted have, in turn, produced certain organizational arrangements that are distinctive in each state.

Property Rights and Control of Groundwater

Broadly speaking, the purposes of conjunctive management in Arizona and California are similar—long-term storage, overdraft recovery, and satisfying the financial and operational needs of large-scale water projects. There are substantial differences, however, in the types and organization of conjunctive management projects in the two states, and those differences are traceable to institutional arrangements, especially groundwater rights.

Beginning in 1980, the State of Arizona recognized, developed, and allocated individual property rights in groundwater in the state's most heavily overused basins. Shortly thereafter Arizona recognized property rights in recharged groundwater as well. California has not developed or allocated a system of property rights in groundwater basins, although if water appropriators within a basin develop their own system of rights through adjudication, courts will generally recognize and enforce them (Blomquist 1992).

This difference alone accounts for an important distinction between conjunctive management projects in the two states. In Arizona, a water appropriator in an Active Management Area (AMA) who wishes to pursue a conjunctive management project on its own can do so (see bottom half of Figure 7-1). This is true of both private and public entities with appropriative rights to groundwater. The appropriator can acquire surplus surface water and store it underground through a direct or in-lieu recharge project, and be assured that the stored water will be available for its own use, or available for sale to another water appropriator. Acting alone, an appropriator can develop its own projects and capture the benefits of doing so.[2]

Because state law in California provides no such security, particularly with respect to storage and recapture, groundwater users do not invest in conjunctive management projects without first investing in institutional arrangements that allow them to control groundwater basins (see bottom half of Figure 7-2). Conjunctive management projects occur only in basins where users have developed governing arrangements that allow them to control access to, withdrawal from, and management of groundwater basins (Heikkila 2001); otherwise, nonparticipants could divert the project's benefits. Water users in California must do for their own basin what the State of Arizona has done for AMAs—define rights of withdrawal, storage, and recapture—before engaging in conjunctive management projects.

The logic of this connection between institutions and conjunctive management practices extends to Colorado. There, well-defined rights in water give appropriators the ability to act on their own in pursuing conjunctive water management projects (see bottom half of Figure 7-3), just as appropriators do in Arizona. The water-rights system assures participants in augmentation and recharge plans of the ability to capture the benefits of their efforts. Thus, neither Colorado nor Arizona appropriators have to gain control of entire basins as California water users do before engaging in conjunctive management, and they may pursue conjunctive management on their own instead.

What is common to the three states is that conjunctive management required institutional arrangements limiting access to and pumping from groundwater basins. This is necessary to assure those who invest in conjunctive management projects that the water they store will not be taken by others. What is different between California on one hand, and Arizona and Colorado on the other, is how those institutional arrangements were forged—basin by basin in the former, or through a change in state water law in the latter.

State laws recognizing rights in recharged water, and, at least in Arizona and Colorado, allowing such rights to be transferred, further facilitate conjunctive management. Transferable water storage rights allow water users to capture the benefits of conjunctive management by building credits that could be used to meet assured water supply requirements in Arizona, or to cover out-of-priority pumping in Colorado, or could be transferred to other water users needing such credits in either state.

The close link between the establishment of assured and limited groundwater rights and the development of conjunctive management is most clearly illustrated in California, where groundwater governance is largely a local affair. Conjunctive management has occurred *only* in basins where water users have devised basinwide governing systems that limit access and regulate water pumping and storage. Conjunctive management does not occur in basins where overlying landowners and appropriators are not limited in their pumping and cannot be assured that they will capture the benefits of pumping restraint.

Recharge Methods

Both primary methods of groundwater replenishment—in-lieu and direct recharge—are used in all three states, but variations do appear. In-lieu recharge may account for the largest volume of water conserved in the three states, in-lieu recharge techniques in Colorado differ substantially from those in Arizona and California. In those two states, in-lieu recharge allows large state or regional project water contractors to supply additional surface water to local water users or suppliers in exchange for reduced groundwater pumping. The state or regional project operator obtains in exchange the option to provide less surface water in dry periods and to have the local entities draw down stored groundwater. Instead of these cross-temporal exchange agreements where water from one source now is swapped for water from the other source later, Colorado's projects involve groundwater users directly replacing depleted surface flows with another water source. This distinctiveness to Colorado's conjunctive management practices is also traceable to institutional arrangements there—because groundwater rights have been integrated into the state's prior appropriation system, junior groundwater users can be "called out" or shut down if stream flows are diminished in ways that harm senior surface water users. Groundwater pumpers therefore acquire rights to supplemental water that can be delivered to the stream on short notice so as to be able to continue pumping in dry conditions.

Direct recharge projects are the most common type of conjunctive management projects in the three states, even though in-lieu arrangements account for more water. Direct recharge projects also differ among the states. While direct recharge projects in all three states rely on percolation ponds or basins, those in Colorado are suited to the purpose of allowing groundwater users to replace depleted stream flows. These constructed ponds or recharge sites projects are located adjacent to rivers to maximize the net return to the surface water rights holders. They also tend to be smaller in scale than recharge operations in Arizona and California, because in Colorado a groundwater user or groundwater users' association is only trying to replace its own depletions and not necessarily manage the resource conditions of the entire basin. In California and Arizona, projects are not geared toward replacing or maintaining stream flows, and therefore recharge sites in those two states may be located at considerable distances from rivers and streams. The siting of recharge operations depends instead on favorable percolation rates, depth to the water table, and the location of project water infrastructure. Most of these programs are directed toward long-term storage and rely on constructed percolation basins for storing project water or treated effluent underground.[3]

Organizational Arrangements in the Practice of Conjunctive Management

Our discussion of differences in the practice of conjunctive management in Arizona, California, and Colorado now leads us to the types and roles of water management organizations. In Chapter 3, we suggested that conjunctive management was likely to be shaped not only by institutional arrangements such as water rights but by organizational structures as well. Here we trace those connections by continuing our comparison of the three states.

Local Organizations and Interorganizational Relationships

In California, the development and implementation of conjunctive management has occurred primarily through the efforts and interaction of three types of organizations—large-scale water project operators, basin-level management authorities, and basin-level water user associations. Large-scale water project operators such as the California Department of Water Resources, the U.S. Bureau of Reclamation, the Metropolitan Water District of Southern California, and in one case the Los Angeles Department of Water and Power have participated in conjunctive management projects largely because of their interest in smoothing out variations in water demands. Of course, these project water sellers have needed willing and capable "buyers." As we have observed, California water-rights law does not encourage a single entity within a groundwater basin to acquire and store surplus water underground because that entity cannot control it. Basin-scale organizations have been needed to perform this function, as California water law discourages any single groundwater user from storing surplus water underground where they cannot control its disposition.

In several of the California basins we studied, pumpers originated basin governance by organizing water-user associations or working through existing local networks such as farm bureaus or chambers of commerce. These private organizations provided forums in which basin users could discuss their concerns and experiences with overdraft and other problems, and explore solutions. Some of the earliest efforts to design groundwater management programs, and hence conjunctive management programs, took place within these self-organized associations.

Such organizations have inherent limitations, however, when it comes to actually implementing conjunctive management projects. Financing and constructing needed infrastructure, assessing water users to repay those costs and to sustain the ongoing operation of one or more projects, monitoring stored water and allocating access to it required entities with public powers. Thus, a third principal type of organization involved in conjunctive management in California has been the basin-scale management authority.

There are several variations of these authorities—water districts, watermasters, replenishment districts, and so forth—but they have in common the responsibility of overseeing and/or carrying out the enterprise of conjunctive management at the basin level. Some evolved from water user associations whose members came to the conclusion that they needed a different organizational form to perform conjunctive management functions, and some of these basin-level management authorities are still advised by water user associations. Since the adoption of state enabling legislation such as Assembly Bill (AB) 3030, discussed in Chapter 4, other types of local governments have been authorized to engage in several of the functions of basin-level management authorities. Regardless of how they came into existence, these basin management authorities interact with the large-scale water purveyors through contracts or other arrangements to coordinate the use of surplus surface water with the use of the basin's underground storage capacity.

What has therefore emerged in California is something like a marketplace approach to conjunctive management. Project operators with occasional water surpluses, or with needs to supplement dry-year yields—especially the California Department of Water Resources and the Metropolitan Water District of Southern California—have searched for buyers who buy and bank surplus water in wet years to be recovered when needed later. Metropolitan in particular has forged dry-year storage contracts of this sort with several local water districts as described in Chapter 4. Similarly, management authorities in basins with available storage capacity, usually as a result of overdrafting, have searched for suppliers with surplus water to store. Some basin organizations have even explored prospects for letting neighboring or distant communities store water in their basins for recapture or exchange when needed, and some communities without underground storage capacity have searched for water-banking arrangements with those that do.

One might expect that in such an environment, innovative institutional arrangements for water storage and transfer would be devised and numerous transactions among mutually self-interested partners would occur. The former proposition has been more in evidence than the latter. Creative arrangements for purchase, storage, recapture, and exchange have indeed been worked out in California, but not nearly as many as would have transpired in an environment in which greater certainty about water rights existed. In California's many basins where pumping rights have not been determined and limited, the lack of assurance concerning the future availability of stored water continues to inhibit the realization of the state's conjunctive management potential.

Basinwide water organizations may not be needed in Arizona and Colorado as they are in California, but other water organizations play important roles. In Arizona, the state has created the Central Arizona Groundwater Replenishment District (CAGRD) and the Arizona Water Banking Authority to encourage wider use of conjunctive management. As described in Chapter 5, CAGRD contracts with residential and municipal water purveyors to replace mined groundwater, as required by state law, with CAP water the district has stored in multiple projects. The Arizona Water Banking Authority (AWBA) arranges with organizations that have recharge projects to store surplus CAP water that would otherwise be lost to the state.

In Colorado, water users have devised private water associations to assist well owners in meeting the requirements of the prior appropriation system. As described in Chapter 6, well-owners' associations in the South Platte and Arkansas River basins reduce the transaction costs of incorporating groundwater within the prior appropriation system for the state and for individual well owners. Instead of searching for, purchasing, and making available to the state sufficient surplus water to cover out-of-priority pumping, a well owner can join an association that acquires quantities of water sufficient to cover all its members. Instead of the state monitoring and interacting with each well owner, it can work with a well-owners' association to ensure compliance with Colorado law.

Local and regional water associations are therefore important to conjunctive management in each state, but they differ by type and play distinct roles as summarized in Table 7-1. It should also be noted that water organizations composed of multiple water users to facilitate conjunctive management often take a decade or more to develop. Organizations emerge slowly as water users sort out their differences and search for common ground. For instance, Arizona water users first experimented with creating basinwide replenishment districts to pool water among users within a single basin. The legislature adopted enabling legislation allowing water users in the Tucson AMA and the Phoenix AMA to create their own replenishment districts. Water users in each AMA could not settle differences over how to fairly allocate the benefits and costs of such districts, and both failed. Smaller water providers who were not offered or did not sign CAP contracts were left in dire straits—facing assured water supply requirements without access to renewable sources of supply. The state eventually responded by creating CAGRD. In California, basin adjudications take decades to complete as numerous and diverse water users gather information about their water activities and problems and search for equitable solutions. Even water users that have avoided adjudication and have instead attempted to create AB 3030 plans find that resolving differences and reaching agreement on how to manage basins is not easy. Many AB 3030 plans have yet to be completed after several years of negotiations. Thus, while water organizations have a vital role to play in supporting conjunctive management, they are neither quick nor easy to create.

The Role of the State Government

Another important element of conjunctive management, and an important point of comparison among the states, has to do with the role of state government in conjunctive management. One aspect of that role relates to the state–local relationship in each state. Another has to do with the roles played by state water agencies in facilitating conjunctive management.

As mentioned briefly in Chapter 1, California and Colorado have long and vibrant traditions of local rule, and Arizona does not. In California and Colorado, local jurisdictions and water users exercise substantial authority in relation to governing water. The state government in Colorado has provided an institutional setting in which water appropriators are encouraged to govern themselves, and even facilitated in doing so. Water appropriators have created their own associations that allow them to act collectively to protect their water rights or to interact with the state. Local water commissioners monitor appropriations. Water appropriators define, transfer, and defend their water rights within the system of water courts maintained by the state. The water courts provide a forum in which appropriators can resolve their conflicts and devise and enforce binding agreements among themselves. For the most part, the Colorado legislature has followed rather than guided these local arrangements, recognizing and codifying into state law the water laws and practices devised in water courts. The involvement of the Colorado Supreme Court and the Legislature during 2003 in phasing out the state engineer's practice of approving replacement plans and substitute supply plans, although it will make conjunctive management projects more costly and time-consuming, nevertheless proceeded from the recognition in Colorado law that the water courts in each hydrologic division of the state were the institutional forum for approving such arrangements. The locus of water governance in Colorado remains at the local level.

Local governance holds true in California as well, where the legislature has accommodated local communities' requests to establish water districts, often on a case-by-case basis to fit local circumstances. One result is the vast number of special districts in the state with diverse sets of powers. The state's willingness to allow local water users to design their own institutions of authority for developing and managing water supplies has encouraged the basin-by-basin approach that characterizes conjunctive management in California.

Even the lack of a statewide system in California for setting rights of groundwater withdrawal, storage, and recapture reflects and reinforces the tradition of local governance. That tradition supports the notion that groundwater problems, as local matters, should be resolved primarily at the local level. This view has justified and sustained the legislature's unwillingness to set down statewide groundwater rules. At the same time, the absence of those statewide rules leads local water users to devise their own basin-scale arrangements, which include the formation of basin-level organizations and the use of basin adjudications. With the exception of the State Water Project—which has been a major source of replenishment water in California conjunctive management projects since 1970—nearly every aspect of conjunctive management in California has been developed and refined through local action.

In Arizona, the state government has taken a much more direct and active role in managing water and in directly developing and participating in water projects. The AMAs are designated by the state, with state-defined boundaries and rules governing water use. As Arizona exhibits fewer locally designed water management organizations and local forums for resolving water conflicts, the state government has facilitated the development and implementation of conjunctive management in a number of ways. We have already noted the roles of CAGRD and AWBA. AWBA and the Central Arizona Water Conservation District, under which CAGRD operates, together held storage permits in nearly 75 percent of the projects operating in 1998. The Central Arizona Water Conservation District even sponsored the facility permits for four of the projects in our study.

Another way Arizona government takes a direct role in conjunctive management is through the administration and monitoring of projects. The legislature created the Arizona Department of Water Resources to implement and administer the 1980 Groundwater Management Act and develop rules for assured water supplies in the AMAs. These rules encourage municipalities and developers to participate in conjunctive management by becoming partners with CAGRD in storing surplus CAP water for long-term credits. The legislature also gave the Department of Water Resources the authority to establish the permit process for groundwater storage, savings, and recovery under the 1986 Act and its pursuant amendments. Thus, state government establishes and monitors short-term and long-term credits for direct and in-lieu groundwater storage projects.

The role of state agencies in California and Colorado is more supportive and facilitative of local efforts, rather than directing and managing conjunctive management projects. The Colorado state engineer's office, for example, performs a number of functions that assist in the development and implementation of conjunctive management projects. The Colorado state engineer's office provides technical support to appropriators and water courts. Recently the state engineer began using rule making authority to assist in incorporating groundwater into the prior appropriation system, although those rules are adjudicated in water court. Division engineers assist appropriators in coordinating their water activities on a basinwide scale. The state and division engineer's offices are invaluable repositories of records of water-rights holders and of authorized augmentation or replacement plans, records that can be essential to the successful operation of a conjunctive management program.

In California, the role of state agencies in facilitating conjunctive management is also limited. The California Department of Water Resources provides a considerable amount of data on state water conditions, much of which is useful to local water organizations in designing conjunctive management projects. The department managed the state's Drought Water Bank in 1990 and 1992, but that was primarily a water transfer arrangement rather than a conjunctive management operation. As noted in the preceding, the State Water Project division within the department does participate in some conjunctive management projects as a water supplier. In addition, as described in Chapter 4, the department's Division of Planning and Local Assistance has been much more active recently in promoting and facilitating the development of local conjunctive management projects, including the distribution of funds from water bonds approved by voters in 2000 and 2002.

On the other hand, California state government does not maintain a common inventory of conjunctive management projects in the state such as one might find in Arizona through the water banking authority or in Colorado through the recording of augmentation and replacement plans. Although the California Water Resources Control Board maintains records of surface water appropriation rights, since there is no statewide system of groundwater rights there is similarly no state agency from which one can obtain a list of groundwater users in any basin. To a great extent, water providers and water managers in California rely on formal and informal networks such as the Groundwater Resources Association of California and the Association of Ground Water Agencies to learn from one another about experiences in devising and implementing conjunctive management projects.

Do state officials support conjunctive management projects from a concern for advancing improved water-resource management? To some degree perhaps, but it is also clear that facilitating conjunctive management satisfies the interests and incentives of state officials in all three states. Arizona officials have been motivated throughout the past half-century to protect Arizona's Colorado River allocation and secure federal funding for the CAP, so the ensuing high degree of involvement of state government in promoting conjunctive management using CAP water for recharge is not surprising. In California, state officials have had an interest in maintaining the operation of the large-scale water projects in the state through an era when federal regulations and public opinion have promoted a wider array of environmental values for water, and in arresting groundwater overdraft problems that have reached acute conditions in the Central Valley and portions of Southern California. Yet few if any state officials seem interested in paying the political price of reforming water-rights law in the state, especially with respect to groundwater. Whatever its other virtues, facilitating local conjunctive management programs can be understood as a way to achieve the water policy goals without paying that price. In Colorado, having paid the price of integrating groundwater into the prior appropriation doctrine, state officials had an interest in avoiding the foreseeable consequences of that action—a complete or nearly complete shutdown of groundwater pumping on which many of the state's newer communities and much of its urban and suburban population, that is, voters and taxpayers, rely. Colorado officials also have to deal with the interstate compacts and federal regulations that require specific amounts of surface water to exit the state. Facilitating conjunctive management projects that allow groundwater pumping to continue, and supplement the flows of surface streams at the state's borders, helps to satisfy those concerns.

Conclusion

Distinctions among the legal and organizational settings of the states translate into opportunities and constraints that have shaped the emergence and practice of conjunctive management. There are some similarities in conjunctive management practices, such as the dominance of in-lieu recharge and the role of large projects as water suppliers, but the differences among the states are more numerous and can be traced directly to their diverse institutional settings.

It also appears that the relationships among institutional rules and water or-ganizations involved with conjunctive management are interactive and dy-namic. Aspects of the institutional setting—such as water-rights laws, or the relationship between state and local government—shape the purposes of con-junctive management and affect the types of organizations water users craft to try to implement it. As water users engage in and learn about conjunctive man-agement, they in turn attempt to change institutional arrangements in ways more supportive of their conjunctive management goals. Certainly Californi-ans have created and adapted basin-scale organizations in response to the bar-riers that the state's water-rights system placed in the way of conjunctive man-agement. Those basin-level entities have experimented with a variety of replenishment water sources and recharge methods over the course of con-junctive management's history in that state so far. Although Arizonans came into the conjunctive management scene later than others, they have modified state statutes governing groundwater storage and recovery in ways that have clearly advanced the goal of expanding conjunctive management, ensuring that credits are available for project participants, and establishing organizational entities such as CAGRD and AWBA. Colorado water users remain fiercely tied to the prior appropriation doctrine, but have worked to reconcile the protec-tion of surface water rights and stream flows with groundwater pumping through rule changes and the formation of water-user associations.

The recognition of the importance of institutions in the water policy liter-ature is accurate: institutions indeed matter. Studying these three states in some depth, and comparing them, has given us the opportunity to examine and clar-ify how they matter. In this and the previous three chapters we have attempted a close and thorough consideration of *how* institutions are related to water management—how they stimulate changes in management practices, how they facilitate or hinder those changes, how they shape the choices water users and organizations make, and how they affect the outcomes water users and or-ganizations achieve.

As we have seen, in some cases institutional arrangements make conjunc-tive management more difficult to achieve. Property rights in water may not provide sufficient security to induce investment in conjunctive management. Water users may find conjunctive management attractive but financially pro-hibitive, and institutional arrangements may impair their efforts to coordinate and finance joint projects. It is possible to identify several instances where in-stitutional arrangements have thwarted conjunctive management efforts.

The obstacles created by institutions are not the whole story, however. Iden-tifying institutions as the problem may be accurate in many instances but it is also incomplete, just as noting that institutions matter is an accurate but in-complete observation. Institutions matter not only because they inhibit, but also because they are absolutely necessary for realizing, desirable water prac-tices. Well designed institutional arrangements support investment in conjunctive management, ease the costs of transferring water to different locales and for different uses, support the coordination efforts of multiple water users, and so forth.

Institutional arrangements are tools with which citizens and officials in the West will address and ease their water resource and use problems. Having explored the past and present use of conjunctive management in Arizona, California, and Colorado, we turn next to the future, and then to some suggestions for institutional changes that would support and extend it in those three states and possibly elsewhere.

8

Future Directions of the Diverging Streams

The western water landscape is never static, but its changes are not random either. Changes occur along paths established by prior choices and actions. Our look to the future of conjunctive management in California, Arizona, and Colorado is therefore based on existing water policies and practices in those states, as well as on broader trends in water conditions and water-resource management.

In looking ahead, we ask about each state: How might conjunctive management evolve, as one aspect of water resource management, over the next few decades? Will water managers take greater advantage of opportunities to improve conjunctive management or struggle along with separate management of ground and surface water supplies? What trends are driving policymakers and water users toward conjunctive management or away from it? How will the institutional arrangements discussed in this book shape their ability to adapt and meet growing and competing water demands? We close the chapter with a discussion of some regional issues potentially affecting all three states.

The Future of Conjunctive Management in California

Water supply crises have often preceded the development of institutions for conjunctive management in California. Although a single event, such as a prolonged drought or a massive earthquake, might produce such a crisis, the diversity of the state's supplies and the extent of its infrastructure make it more likely that water supply crises will arise from a convergence of trends. There is no shortage of such trends in California, however, and significant changes are on the horizon. The future of conjunctive management and other water management efforts will depend on the institutional responses of California policymakers to current and expected water management dilemmas.

Several trends are intensifying interest in the expanded use of conjunctive management in California. In brief, they are: (1) changes in the operation of large-scale water projects, (2) increased stringency of assured water supply requirements, (3) river restoration movements, and (4) the development of water reuse as an alternative water source. Each trend is joined to institutional features that make conjunctive management an attractive option.

Changes in Large-Scale Water Projects

Foremost among the trends in California drawing attention to conjunctive management and other water management options are the altered operations of the large water projects that supply most of the population and much of irrigated agriculture. As noted in Chapter 2, some of the annual yield of the federal Central Valley Project has been redirected already to environmental needs in the San Francisco Bay/Sacramento–San Joaquin Delta region. Similar prospects are in store for the State Water Project as an outcome of the CALFED process attempting to find long-term solutions to Bay–Delta water quality issues. Although successful conclusion of the CALFED seems to be continually pushed off the horizon by new issues and problems,[1] whatever changes are made to aid the Bay–Delta ecosystem are likely to affect both the timing and the quantity of water deliveries through the State Project.

New developments on the Colorado River affect the largest other water importation project. In exchange for interim surplus operating criteria on the Colorado that allows a transition period of several years, California has agreed to reduce its draw on the Colorado River in nonsurplus years to the 4.4 million acre-feet allocated to it in the Colorado River Compact. Limiting Colorado River imports to 4.4 million acre-feet represents a reduction of nearly a million acre-feet per year, most of which was relied on by the Metropolitan Water District of Southern California (MWD) and its population of nearly 17 million Californians. An agreement reached in October 2003 allowed for some shifting of Colorado River supplies among California water agencies to soften the impact of the reductions on MWD and its member agencies, especially San Diego County Water Authority (Perry 2003).

Changes in the priority allocation of water from the State Water Project and the Colorado River Aqueduct will be felt principally in dry years. Alternative sources of supply for agricultural and urban consumptive water uses will be needed along with conservation, but will not be enough because the issue is timing of supplies and not only quantity. Reliably meeting consumptive water uses as well as environmental needs will require more storage capacity to bank water during wetter periods. Some of that additional capacity may consist of surface storage, and several projects are already being designed, and state and federal funds sought, to increase the heights of some existing dams, and even to site one or two new reservoirs. But storing water underground could be less expensive, as well as less politically challenging.

Because MWD and its member agencies receive water from both the State Water Project and the Colorado River Aqueduct, prospective reductions in deliveries from those sources affect them the most. Conjunctive management has thus been one of MWD's top priorities in current practices and future plans. As noted in Chapter 4, MWD has been actively seeking dry-year transfers and water storage arrangements, both within and beyond its service area. Along its Colorado River aqueduct route, MWD has identified groundwater basins where it would recharge surplus water from the Colorado in wet years, to draw on in dry years to meet its delivery obligations to member agencies. Along the route of the State Water Project, MWD has identified basins where wet-year supplies might be stored if it did not need its full State Project allocation within that year. The district can be expected to continue, and perhaps even intensify those efforts in the future while it also supports local agencies' efforts to develop new supply sources from desalination and the purification and reuse of wastewater.

The California Department of Water Resources—operator of the State Water Project and a key participant in the CALFED process to address the Bay–Delta problems—is also exploring conjunctive management and other project opportunities that would enhance dry-year options and accommodate the reoperation of the State Project. The department's search has concentrated on sites in the Sacramento River watershed, where water banked from the Sacramento and tributary rivers during surplus periods could be returned to the river, and thus to the Bay–Delta region, in dry periods.

MWD and Department of Water Resources (DWR) staff see conjunctive management as a comparatively low-cost and low-impact way of addressing some of the changes in the state's water supply situation that are clearly approaching at an accelerating pace. In their searches for new conjunctive management and water transfer prospects, MWD and DWR have understandably focused on and expressed their preferences for arranging a few large-scale projects rather than numerous small ones. They would perhaps be inclined to develop some conjunctive management projects on the scale of the larger ones in Arizona. But in California, the larger deals are the most difficult ones to complete. Political conflict accompanies any large water deal in California—urban, agricultural, and environmental constituencies contending over whether a project will secure stable supplies at reasonable costs for urban uses, take water from agriculture or inadequately compensate farmers for the impacts to them, fuel further growth that aggravates the state's water problems in the future, and so on. In addition, California's local, basin-by-basin approach to groundwater management compounds the number of individuals and organizations whose cooperation must be secured and whose actions must be coordinated. These transaction costs also make larger projects harder to accomplish than smaller ones. Hence, although changes in the operation of large-scale water projects intensify the perceived needs and advantages of conjunctive management, it remains unclear that new large-scale projects will result.

Assured Water-Supply Requirements

Another trend affecting the future of water management, including conjunctive management, is increased stringency of assured water supply requirements. After six years of wetter than average conditions, Californians faced drought again in 2001 and 2002. These conditions, together with other factors, prompted California legislators and judges to move toward assured water supply requirements for new development in the state.

In 2001, the legislature passed and the governor signed two bills requiring demonstration of assured water supplies for large new developments. One tightened a requirement at the permit application stage that was already on the books but being ignored in practice. Chapter 643 of the Statutes of 2001 was intended to tighten the requirements for information provided in environmental impact reports (EIRs) and in state-mandated Urban Water Management Plans (UWMPs). EIRs, which identify the public water systems that will be used to supply water for a new development, will have to identify the sources of additional water more specifically, describing the water supply entitlements, such as contracts and water rights, held by that public water system, and any projects that will be necessary to meet the needs of the new development. UWMPs, which are submitted every five years by water providers in urban areas, will have to include descriptions of all water-supply projects and programs that will be needed to meet projected needs in the provider's service area, including projects to develop groundwater when it has been identified as a supply source for that area.

The other statute, Chapter 642, imposed a new obligation on developers and local governments granting final preconstruction approval to account for the water supplies to support a development. This law reaches beyond California water agencies and providers, and affects general-purpose governments such as counties with authority over land use. Those governments may not approve a subdivision map, parcel map, or development agreement for a development of more than 500 dwelling units without first obtaining from the applicable public water system a written verification that adequate water supplies are currently available for the new development, or will be available before the project is completed (Robie 2001).

Action in the courts paralleled that of the legislature. Also in 2001, a California appellate court decided in 2001 that an EIR was inadequate because it failed to identify sources of water supply and means of wastewater treatment and disposal for future developments foreseen in amendments to Napa County's general plan.[2] The question of adequate showing of water supplies to support new development also has had large proposed communities such as Newhall Ranch and Ahmanson Ranch in Los Angeles County in and out of courts through 2000–2003.

Institutional changes such as these, combined with the changes in large-scale water projects discussed previously, provide further impetus toward expanded use of alternative water supplies and toward greater use of conjunctive management. In the past, limitations on water availability for new development in California were often overcome by acquiring supplemental water supplies from the operators of imported water projects. That option seems less promising now, to say the least, and at most it may be foreclosed altogether. So in this new setting, if assured water supply requirements are implemented in California as they have been in the Active Management Areas (AMAs) of Arizona or the Arkansas River Basin in Colorado, developers and local governments wishing to support development and needing to enhance their portfolio of reliable water source will have new incentives to enter into water banking, storage, and transfer arrangements.

River Restoration

Another factor shaping the future of conjunctive management in California is the relatively sudden but nonetheless widely popular movement for river and stream restoration in California. Although this movement understandably draws gibes and chuckles from outside observers in light of California's reputation for doing almost anything with rivers and streams except letting them flow in their natural channels, several local agencies have been formed—here too we see California's preference for organizing at the local level—and hundreds of millions of dollars of state funding have been committed to this purpose. River restoration—especially urban rivers and streams, which seem to be the focus of much of the organizing and popular support—is increasing the emphasis on the preservation and maintenance of stream flows. In light of California's climatological variability, maintaining stream flows for river restoration while also meeting consumptive water demands is likely to involve some means of banking water when it is plentiful in order to supplement flows when needed.

Water Reuse

Further development of water reuse in California also holds potential for increased attention to conjunctive management. The connection between these two aspects of water resource management is not obvious; it arises from the state's regulations governing the uses of reclaimed water.

For both regulatory and political reasons, water reuse cannot be direct, even when the water has been purified to a quality superior to drinking water standards. (The political death phrase is "toilet to tap.") Reclaimed water can be introduced into public water supplies only indirectly. Indirect potable reuse involves mixing the treated water with other sources first, and allowing a substantial residence period prior to introducing the water into public supplies. This is most commonly accomplished, at least in California, through recharge into an aquifer where the treated water mixes with other groundwater and resides for a determined period before being pumped out and reused.

Conjunctive management—recharge, store, and recapture—is thus a key to fulfilling the potential of water reuse in California. But here too institutional obstacles are encountered, despite general policy pronouncements by state officials favoring reuse as an environmentally and economically smart option. Substantial regulatory restrictions exist even on indirect potable reuse, and California Department of Health Services approval for aquifer recharge with treated water has been a highly uncertain proposition. A new era may have opened in October 2003 with the approval of the largest project to date for groundwater recharge with reclaimed water,[3] but it is too early to draw any conclusions.

These trends in California favor the likely development of more conjunctive management in the near future. How it will evolve is not as clear. California may have the longest history of conjunctive management of the states we have studied, but it is also the most geographically expansive and diverse, most populous, and most institutionally complex, all of which make the future of conjunctive management harder to forecast. Pursuit of conjunctive management in California has always been an uphill struggle impeded by institutional obstacles such as the state's water-rights rules. Without corresponding changes in the institutional impediments to conjunctive management, it remains uncertain whether and how Californians will employ conjunctive management to respond to the water management issues facing the state.

The Future of Conjunctive Management in Arizona

In Arizona as in California, finding more efficient means to manage scarce water resources is a recurring and fervently debated political topic. Perhaps to a greater degree than in California, however, conjunctive management is increasingly recognized as a viable solution to Arizona's water supply problems.

In the summer of 2000, Governor Hull appointed a Commission to review the Arizona Groundwater Management Act and propose legislative changes to help Arizonans maintain reliable water supplies into the future. The commission's recommendations were issued in December 2001, and the Arizona legislature took up water supply management as a major issue during 2002. Most of the Commission's recommendations dealt at least indirectly with aspects of conjunctive management (Governor's Water Management Commission 2001). One proposal was to require the Central Arizona Groundwater Replenishment District (CAGRD) to establish a replenishment reserve to help secure water supplies for CAGRD members; the district was also authorized to impose an enrollment fee for new subdivisions. The foreseeable impact of such efforts in Arizona is that conjunctive management programs will grow.

Another reason to expect further expansion of conjunctive management in Arizona is that the state will continue to have excess Colorado River water supplies over the next two decades, and as noted in Chapter 7, long-term storage of those surpluses is one of the main purposes of conjunctive management in Arizona. The Arizona Water Banking Authority (AWBA), for instance, estimates that it should be able to store at least 8 million acre-feet of excess Central Arizona Project (CAP) water over the next 20 years (AWBA 2000).

Over that period, two trends appear on the horizon in Arizona: (1) increased use of conjunctive management for municipal storage but reduced reliance on conjunctive management for agricultural water use and (2) directed conjunctive management programs for maintaining or improving the condition of groundwater supplies. These trends depend on continued access to Colorado River surpluses. Then, as Arizonans begin to consume their full share of Colorado River water, two other trends in conjunctive management are possible: (1) increased reliance on effluent and (2) increased use of seasonal programs for underground storage. Each of these potential outcomes is discussed in greater detail below.

Long-Term Municipal Storage

Urban water demand in Arizona is expected to continue to rise over the next 25 years, but agricultural water demand is likely to decline as farmlands in Arizona are being taken over by development. Conjunctive management will therefore be an important tool for growing urban areas trying to meet assured water supply requirements. Indeed, a number of new projects have emerged in Arizona's largest cities since the late 1990s. By the end of 2001, 66 projects had been permitted in Arizona [Arizona Department of Water Resources (ADWR) 2001c], an increase of 24 from the total identified in our study three years earlier. Municipalities and developers in the Phoenix metropolitan area manage most of these new projects.

Because much of the water currently going into conjunctive management comes from irrigation districts taking CAP water in lieu of pumping, however, the pace of water storage in conjunctive management programs could slow as agricultural lands become developed. Some smaller irrigation districts with permits from the early 1990s are already reducing their in-lieu water storage programs because of development within district boundaries. Municipal water consumption is far lower per acre than agricultural uses, so some deceleration of in-lieu groundwater savings is to be expected and will not necessarily hurt conservation or long-term storage efforts.

The use of conjunctive management in the growing suburbs of the Tucson metropolitan area will probably increase at a slower pace than in Phoenix, despite Tucson's heavy reliance on groundwater and serious basin subsidence problems. The Tucson AMA faces physical limitations and a lack of infrastructure that may inhibit the growth of conjunctive management. The Tucson AMA office has recognized that future implementation of conjunctive use projects is limited by problems of conveyance, potential contamination sources, and local regulations (ADWR 1999b). The lack of conveyance facilities for some of the smaller but fast-growing communities surrounding Tucson is especially problematic. Many of these communities lack the financial resources to bring capital-intensive CAP canals to areas with good recharge capacity.[4]

Groundwater Stabilization: Storage and Recovery Issues

Although Arizona has moved rapidly into conjunctive management and stored large quantities of water underground, the basins in which water has been stored have not necessarily been located in the most heavily overdrafted and groundwater-dependent areas of the state. As more projects are established in growing urban areas, conjunctive management will rise as a means of addressing groundwater overdraft problems. The Phoenix AMA, for instance, has proposed that the AWBA consider locating some storage programs in rapidly growing areas that face serious groundwater level declines, such as North Phoenix and Scottsdale or near Luke Air Force Base (ADWR 1999a). By storing excess CAP supplies in these areas, and then relying on the AWBA's credit system to recover supplies from less vulnerable areas of the basin, conjunctive management may be used to help stabilize declining groundwater tables.

One of the key questions about the future of conjunctive management in Arizona is the location of future recovery of stored groundwater. Most recharged water in Arizona has not yet been recovered.[5] Currently, credits can be transferred across entities in the same AMA, and recovery well permits for entities storing water are not tied physically to recharge sites. The recovery of water from conjunctive management projects therefore does not have to occur in the place where the water was stored: recovery and recharge simply must occur within the same AMA. A future policy issue facing Arizona is that groundwater overdraft may be exacerbated when a provider stores water in one part of a basin but exercises its stored-water credits by using wells some distance away.

The question of how storage credits will be recovered depends largely on decisions that will be made within the key state agencies. AWBA and the Central Arizona Water Conservation District (CAWCD) hold the largest portion of the state's credits. AWBA is obligated to transfer credits to CAWCD to meet CAP contractors' needs in times of shortage, but other options may be available. An AWBA study commission (1998) recommended that AWBA establish policies for a credit loan program to municipal and industrial users, whereby users would either repay water withdrawals with credits or pay the cost of credit replacement. The study commission also suggested that at some future point it may be useful to give AWBA the power to create a credit pool that could be used when low-priority CAP allocations are unavailable.

The Governor's Groundwater Management Commission mentioned earlier also recommended revisions of Arizona statutes to clarify that the AWBA should begin transferring its long-term storage credits to CAWCD on a yearly basis to provide more assurance of water availability for CAP contractors. Institutional choices such as these on recovery of stored water are still in their infancy, and it remains to be seen whether state agencies will encourage the stabilization of water supplies in areas where it is most needed.[6]

Interstate Water Banking

AWBA is allowed to contract with entities from surrounding states to store surplus Colorado River water. Arizona's experience with large-scale water banking via conjunctive management is now being extended to assist its neighbors. Negotiations with California have been underway for a few years, and those with Nevada recently resulted in an agreement. Here too, institutional arrangements—in this case, the "law of the river" governing apportionment of the Colorado—are driving the development of a new form of conjunctive management project.

Growth in southern Nevada has outpaced that state's apportionment of Colorado River water; indeed, in 2003 Nevada formally requested an increase of 20 percent in its Colorado River allotment, currently fixed at 300,000 acre-feet per year. Even if such an increase were granted, Nevada's growth necessitates other water management measures. On July 3, 2001, agencies in Nevada and Arizona signed an agreement to store some of Arizona's unused Colorado River apportionment now, and allow Nevada to draw upon it later. AWBA, operating in coordination with CAWCD and the Arizona DWR, will capture and store underground up to 1.2 million acre-feet of Colorado River water as it is determined to be available. The Southern Nevada Water Authority will be able to recover up to 100,000 acre-feet per year of stored water, which provides a cushion above its current 300,000 acre-feet per year Colorado River entitlement. Several additional provisions limit the amounts stored and recovered annually under different conditions (Johnson and Gallogly 2001). In light of the challenges both California and Nevada face trying to live within their Colorado River apportionments, such interstate water banking agreements could figure prominently in the future of conjunctive management in the region.

Longer-Term Trends: Water Reuse and Seasonal Storage Projects

Looking somewhat further into the future, two other foreseeable trends in Arizona may be increased use of seasonal storage and recovery projects once Arizonans begin to consume their full share of Colorado River water supplies and increased water reuse as an element in conjunctive management projects. Instead of using surplus Colorado River supplies to store underground and earn long-term storage credits, municipalities and developers may begin to rely more heavily on projects that capture seasonal surplus flows in the winter and spring and then store those flows underground to reuse during drier months—

in other words, to develop some of the kinds of conjunctive management projects found more commonly in California and Colorado.

Likewise, using treated wastewater—more commonly called effluent in Arizona—for short-term storage and recovery could increase. Tucson and a few municipalities around Phoenix have effluent storage projects. Most use the stored water for turf irrigation on golf courses, city parks, or schools. With the state's rising population there will be a need for more of these projects, particularly because effluent may be the only new source of water available.

Thus, existing use of conjunctive management in Arizona follows a fairly clear pattern built around the long-term storage of Colorado River supplies that the state is entitled to but does not need yet. Once excess Colorado River supplies are exhausted, however, the future of conjunctive management in Arizona will be more uncertain. Simply continuing to rely on excess flows of Colorado River water to drop underground for future use will not provide Arizonans with water management security in 2020 and beyond. Renewable water supplies in Arizona eventually will be very limited, and adjustments to water management institutions and their conjunctive management operations will be needed to address these limitations.

The Future of Conjunctive Management in Colorado

Colorado was one of the states most severely affected by the 2000–2002 drought in the West. In April 2002, Colorado's Governor Bill Owens requested that the U.S. Department of Agriculture designate the entire state under emergency drought conditions. As stream flow conditions worsened, water managers in Colorado began to push for stricter enforcement of stream flow replacement obligations and thus for employing Colorado's version of conjunctive management to meet these obligations. In the near future, conjunctive management in Colorado may continue to expand within that pattern. Replacing surface water depletions caused by well pumping is likely to be the primary goal of conjunctive management into the foreseeable future.

What will change is that many of the currently short-term conjunctive management projects in the state may require longer-term water supply assurances, especially for irrigators. Second, although underground storage of surplus surface water in underground basins will probably remain uncommon, it is likely to become increasingly attractive to rapidly growing Front Range cities that desperately need additional water supplies and greater reliability.

Stream Flow Maintenance

The predominant practice of conjunctive management in Colorado is short-term and seasonal—replacing or supplementing stream flows in a dry season with water acquired or stored during a wet season. A significant portion of conjunctive management occurring in the South Platte River and Arkansas River watersheds is built on a "temporary" foundation. Well-owners' associations acquire short-term leases of surface water rights to cover their members' pumping under substitute supply plans and replacement plans.

What could change in the future is the short-term aspect of this practice. First, these short-term plans have not been adjudicated and decreed by a water court, but have been approved annually by the state engineer. Partly because of this practice of approving temporary arrangements, in the South Platte Basin "there is no clear policy governing the amount of replacement water that is needed" (MacDonnell 1988, *592*).[7] As seen in Chapter 6, recent decisions by the Colorado Supreme Court and Colorado Legislature are bringing this practice to an end, and requiring all such arrangements for replacing depleted stream flows and protecting senior surface rights holders to go through the adjudication process and to be decreed as augmentation plans.

This new requirement may well increase the requirements for replacement water, if recent trends in the state engineer's approvals are any guide. Over the past three decades in the South Platte Basin, the state engineer has required the Groundwater Appropriators of the South Platte (GASP) and Central Colorado Water Conservancy District (CCWCD), the two major well associations with substitute supply plans, to increase the amount of water that they must make available to cover well pumping. Initially, the associations provided an amount of water equal to 5 percent of the water their members pumped. By the mid-1990s, the associations provided water equal to 30 percent of pumped groundwater, and the figure continues to rise partly as a result of the 2000–2002 drought and partly due to declining transmountain water project flows. Front Range cities are capturing and reusing transmountain project water to a greater degree, rather than disposing of it in the South Platte River. Their disposal of those transmountain flows—essentially, their discharge of treated wastewater to the river—had served in the past to mask the full effects of groundwater pumping on the river.

In response to increasingly strict replacement requirements, GASP and the state engineer's office have encouraged and worked with farmers, irrigation districts, ditch companies, and towns to develop additional direct recharge projects to cover well pumping along the lower end of the South Platte River. For instance, the Julesburg Irrigation District has developed 21 direct recharge sites as well as using segments of its ditch for recharge (Colorado Stream Lines 2001a). Numerous other ditches have followed suit. In addition, several recharge sites have been added to Tamarack Ranch. No longer is recharge occurring primarily among the districts and ditches in and around Fort Morgan. Instead, the center of recharge is moving east, closer to the Nebraska border. These recharge projects not only more adequately cover well pumping, but they also allow Colorado to better meet its South Platte River compact requirements and its requirements to provide additional water to the South Platte River to recover endangered species. The development of direct recharge projects will continue to be an important means of covering well pumping in the South Platte basin for the foreseeable future.

In the Arkansas River basin, well pumping will continue to be covered through indirect recharge projects with well-owners' associations leasing project water. The transaction costs of leasing surface water may be eased, however, through the creation of a water bank. Legislation passed in 2001 directs the state engineer to promulgate rules for a pilot water bank (Colorado Stream Lines 2001b). Individuals or organizations holding storage shares in reservoirs and willing to lease them would make them available to the bank, which would find willing purchasers. The purpose of the bank is to ease water exchanges without adversely impacting agriculture by intensifying one source of pressure on farmers to sell their water rights (Porter 2001).

In both the South Platte and Arkansas River basins the paths initially established to incorporate groundwater within the prior appropriation system no doubt will continue to be followed. Direct recharge of South Platte River water during low-demand times with the water seeping back to the river during high-demand times, and leasing of project water in the Arkansas River basin will continue for some time to be the dominant forms of conjunctive water management in Colorado.

Long-Term Municipal Storage

Farmers are not the only water users concerned about the future security of water supplies in Colorado. The rapidly growing towns and cities south of Denver in Douglas and Arapahoe counties rely on the Denver basin, which is nontributary, for 85 percent to 90 percent of their water supply (Rouse 2001). Water levels in some areas are dropping almost 30 feet per year, forcing water suppliers to build new wells and increase the capacity of existing wells. The expense of expanding well capacity is mounting rapidly, and cities are searching for renewable supplies of water.

Those searches have focused primarily on tying into existing surface water projects, expanding existing projects, and purchasing agricultural surface water rights. As cities develop additional sources of surface water it is unlikely that such water will be placed in long-term underground storage. Rather it is more likely to become part of indirect recharge activities, allowing cities to switch between surface water and groundwater, depending on their relative abundance. Given the decades required to plan for and develop surplus water sources, this form of conjunctive management is likely to emerge slowly over time.

Some towns are beginning to investigate and use direct recharge to aquifers, however. Centennial Water and Sanitation District, for instance, which serves the Highlands Ranch area, injects treated surface water into the Denver Basin (Rouse 2001). Some cities, such as Aurora, have purchased ranches in adjoining counties with the intention of using the underlying groundwater basins as reservoirs, drawing on them during dry years and replenishing them during wet years. These municipal plans for direct recharge have not yet demonstrated that other rights holders downstream of the ranches will not be harmed (Rouse 2001), and the passage of considerable time may be required to establish whether harm does or does not occur. For reasons such as these, plus the scarcity of surface water supplies that can be stored at all, the long-term underground storage of surface water is unlikely to be widely practiced in Colorado for quite some time.

The irony of the future of conjunctive management in Colorado is that the state's strict adherence to the prior appropriation doctrine pressures tributary groundwater users to engage in conjunctive management but sharply restricts the extent and form of conjunctive management. Tributary groundwater may be pumped, but only to the extent that this does not harm senior surface water rights holders. Out-of-priority pumping may occur only when surplus surface water is available to replace it. Thus, conjunctive management is critical to continued groundwater use in Colorado. Yet the seniority of surface rights combined with the close hydrologic connection between the state's streams and aquifers limits rather strictly the active use and management of groundwater basins. Drawing down basins during droughts, for example, would negatively affect stream flows and thus encroach on surface water users' rights, even if the basins were replenished during times of water abundance. The prior appropriation system as it has evolved and is enforced in Colorado limits more active management of groundwater basins and the millions of acre-feet of water they hold, while at the same time promoting the use of conjunctive management as a means of allowing groundwater pumping. Further development of conjunctive management in Colorado will depend on actions taken by state policymakers to resolve this tension between the law's incentives and the law's requirements.

Regional Trends and Their Implications

Although each state is distinct, the location of California, Colorado, and Arizona in the American Southwest gives them much in common with respect to water resources. Some foreseeable occurrences affecting the whole region can be expected to influence the future development and application of conjunctive management there. From most obvious to least, they include: continued population growth and economic transformation, emerging issues on the Colorado River, and climatic variability and change.

Growth

From the opening page, the extent and rate of growth in the western United States has been a recurring theme in this book. The 2003 California Water Plan

Update is expected to project a 50 percent increase in the state's population by 2030, for example, but only a 10 percent increase in available water supplies (Water Tech Online 2003). Whether one views conjunctive management as a responsible way of addressing problems caused by growth, or as yet another western water management tool that will allow more growth to occur, there can be little doubt that further growth has implications for the entire region and for California, Arizona, and Colorado in particular.

If predictions hold and by the year 2020 or shortly thereafter California contains 50 million residents and Arizona and Colorado surpass 10 million, we can anticipate a number of effects relevant to conjunctive management. First and most plainly, the need for water storage will grow as a means of smoothing out fluctuations in water demands and correcting imbalances between demands and supplies. Second, stresses on surface water supplies will worsen from their already difficult state, even if regulatory protections of species and habitat remain only at current levels of stringency. Third, most clearly in Arizona and probably in California too, some reduction in irrigated agriculture will eventually occur, if for no other reason than the displacement of farmland. Fourth, the economic engine of the region will continue to expand at a rate that outstrips the rest of the country and much of the rest of the world.

These trends have mixed implications for conjunctive management. The first two—increased need for water storage and increased competition among surface water uses—point toward greater reliance on conjunctive management. But reductions in irrigated agriculture may not promote more conjunctive management, for at least three reasons: (1) agricultural water users have been among the principal participants in in-lieu replenishment, water banking, and other variations of conjunctive management projects, (2) groundwater recharge becomes more challenging and more expensive as permeable land is paved for urban development, and (3) reductions in irrigation use may free up water supplies for transfers to urban uses, taking some of the pressure off the development of other techniques such as conjunctive management. Finally, the additional wealth of the region may work in couple of directions: on the one hand, in Arizona, California, and Colorado, increased wealth has often provided the financial wherewithal to meet the costs of initiating conjunctive management projects; on the other hand, increased wealth may facilitate other efforts such as desalination, reuse, and water-use efficiency improvements, which could reduce pressures to pursue conjunctive management.

Emerging Issues on the Colorado River

We have already mentioned implications for conjunctive management flowing from some Colorado River issues: the requirement that California return to its 4.4 million acre-feet annual allotment, Nevada's difficulty with its 300,000 acre-feet annual allotment and request for an increase, and Arizona's banking of its current Colorado River surplus. Recently, noted author Philip Fradkin (2003), commenting on the agreement mentioned earlier among Southern California water agencies for reallocating the state's share of Colorado River water, looked ahead to the future of the Colorado River Basin and observed:

> Today, all the traditional interest groups remain in place and are ready to pounce upon each other more ferociously and with greater desperation than before as the Colorado dwindles . . . the agricultural and urban interests within Southern California; the same division in Northern California, to which the south will inevitably turn again for more water; Indian tribes within California and elsewhere in the West that feel short-changed and are now capable of hiring very effective lawyers with their casino profits; the six other states within the basin that have ganged up on California but have their own internal conflicts, such as the urban-rural split between the east and west slopes in Colorado and the Wasatch Front and the Uintah basin in Utah; and Mexico, whose share is coveted by the seven states and was only grudgingly handed over in the past.

Long-standing struggles in the southwestern United States over scarcity on the Colorado may thus continue or intensify in the future. At least two other issues involving the Colorado River can affect the whole region as well: efforts to save and restore endangered species at the river's end and the possibility of action by the Mexican government to press for dependable deliveries of the Colorado River flows to which it is entitled by international agreement.

During the twentieth century, the flow reaching the Colorado River Delta in Mexico, or the river's mouth in the Gulf of California, diminished to a trickle. In many years, the river no longer reaches all the way to the sea, its moisture swallowed in the sandy soils on the Mexican side of the international border. And the flow that does reach the United States–Mexican border or the river's mouth is high in salts picked up along the course, its quality so impaired as to be of little use for Mexican residents or for species that used to flourish in the delta region. A coalition of environmental groups—Defenders of Wildlife, Environmental Defense, the Sierra Club, the Center for Biological Diversity, Southwest Rivers, and the Pacific Institute for Studies in Development, Environment, and Security—wrote formal protests to the U.S. Secretary of the Interior in 2000, and in June of that year filed a lawsuit in federal district court against the U.S. Secretary of the Interior and four federal agencies challenging the interim surplus guidelines on the Colorado River. Those guidelines, mentioned above, allow the states of Arizona, California, and Nevada to draw more than their fixed apportionment from the Colorado in years when the Interior Secretary declares that surplus conditions exist on the river. Under the guidelines, the Secretary can declare the existence of a surplus even in dry years, by authorizing releases of stored water from Lake Mead. The environmental groups argued that the practice of declaring surpluses even in water-scarce years, and allowing the states to take more than their allotted shares, has contributed to the loss of 95 percent of the wetlands that used to exist in the Colorado River Delta and endangered some species of birds and fish there (Postel et al. 1998; Weissenstein 2000).

That legal challenge to the surplus guidelines did not prevail, as the district court entered summary judgment for the defendants in March 2003, but the issue of endangered species in the Colorado River Delta is unlikely to go away. Should future litigation or a regulatory agency decision end up ordering that adequate flows reach the delta to restore some of the wetlands and protect one or more species there, adjustments will be needed to the operations on the river. At a minimum, Arizona, California, and Nevada could have to modify how much they withdraw from the river and when or under what conditions. Even upstream states such as Colorado could be affected if they were required to allow more water to flow from the upper basin into Lake Mead, where it could be used to regulate flows in and through the lower basin. This could affect some of the transmountain diversion projects in Colorado that move water from the Western Slope to the Front Range, for example.

The most likely consequence for Arizona and California of changes in the operation of the river to meet ecological needs would be a change in the timing of river diversions. The states could be required to concentrate diversions in higher-flow years, and in high-flow months within each year, in order to stay off the river during low-flow periods and allow water to pass downstream. Such timing adjustments will necessitate some regulatory storage to hold the diverted water until it is needed for people or crops. Storing the water underground would be a likely means of handling this adjustment.

The flow of the Colorado River across the international border with Mexico is governed in part by a 1944 treaty committing the United States to ensure annual flows averaging 1.5 million acre-feet per year into Mexico, and rising to 1.7 million acre-feet in surplus years. A 1973 addendum to the treaty established limits for the salt content of the water reaching Mexico, in order to address rising concerns about poor water quality (Morrison et al. 1996, 3–4).

It is plain enough that Colorado River flows have fallen short, in both quantity and quality, of meeting the treaty obligations. Mexican officials have raised the issue from time to time, although neither country approaches the topic with entirely clean hands. Another binational agreement governs the Rio Grande, and Mexico has owed the United States hundreds of thousands of acre-feet in Rio Grande flows. Those flows have been short because development on the Mexican side of that river has not allowed sufficient tributary flow to reach the river itself. The Mexican debt on the Rio Grande and the U.S. debt on the Colorado have produced something of a stalemate, in which neither nation pressed the other too hard.

But the drought conditions of 2001 and 2002 have prompted the United States government and Texas officials to begin pushing for action by Mexico to restore flows on the Rio Grande. It would not be surprising if Mexico attempted to leverage its difficult position on the Rio Grande into a demand for reciprocal behavior from the United States on the Colorado. Should that occur, and if the two nations were to resolve the new dilemma by attempting to live up to their agreements regarding both rivers, the implications for Colorado River users on the U.S. side would be very great indeed. As with the potential changes accompanying species and habitat protection, the most likely modification to the operation of the river would be changes in the timing of withdrawals. This too would likely require the additional use of storage capacity off the river in states such as Arizona and California, which could translate into further reliance on conjunctive management.

Climatological Variability and Climate Change

Still fairly uncertain are the effects of climate change on the southwestern United States, and water managers need more information on the region's climatological variability as well as any long-term future trends (Morehouse 2002; Woodhouse 2002). At a minimum, however, we can look ahead to the potential outcomes of some of the changes that have been forecast.

A continuation of the recent rise in surface and atmospheric temperatures will have consequences for the water supply of the West generally and of Arizona, California, and Colorado in particular. All three states garner a substantial amount of their surface water supply through the year from the melting snowpack that had accumulated the previous winter. Even Arizona, where snowfall is less plentiful, benefits from Colorado River flows that are produced in part by snowmelt from upper basin states. Even modestly higher average temperatures would be associated with a reduced density of the winter snowpack, a reduced area covered by the snowpack because snow elevations on the mountains would rise, and a faster rate of melting in the spring and summer. This assumes that precipitation quantities would stay at their current annual averages, which is also uncertain.

These changes would produce an intensified concentration of runoff and streamflows in the winter and early spring months, increased frequency and severity of flood events during those months, and diminished stream flows in the summer and fall. In short, climate change could exacerbate the variability of surface water supplies and the mismatched timing between peak supplies and peak demands that already challenge western water planners and users. Responding successfully to such changes will require even greater effort to capture runoff and stream flows, protect the region from flooding, and stretch the stored water supplies over longer and drier periods of the year. These too are challenges that may contribute to wider application of conjunctive management.

Although these features are foreseeable, the future of water resource management is always contingent; it depends upon a variety of factors—physical, historical, technological, economic, and especially institutional. It is through the development and alteration of institutional arrangements that people in social settings seek to resolve their shared problems and realize the futures they envision. That is how it is likely to be with conjunctive management in the challenging setting of these rapidly changing southwestern states. To conclude this book, we offer some practical policy recommendations that may guide this evolution of conjunctive management practices and promote more efficient and sustainable use of precious water supplies.

9

Shaping the Future:
Institutional Changes to
Improve Water Management

*Policy analysts who would recommend a single prescription
for common problems have paid little attention to how
diverse institutional arrangements operate in practice.*
—Elinor Ostrom (1998)

As the water-resource conditions and problems affecting Arizona, California, Colorado, and the other regions of the West continue to change, so will the institutional arrangements people craft in response to them. In this concluding chapter we assess the likely effects of some common policy proposals on improved water management practices, particularly conjunctive management. In addition, we offer recommendations for California, Arizona, and Colorado that will facilitate the development of more, and more efficient, conjunctive management programs.

Better conjunctive management will make more water available to western states and communities. Some fear that additional water supplies will simply be used to fuel urban sprawl, and argue that supply-enhancing policies such as conjunctive management should be avoided. We certainly understand that view, but ours is somewhat more positive. Conjunctive management supports growth, but it also allows for the protection of environmentally sensitive and valuable habitats and ecosystems. Assuming, as we do, that the West will experience a larger degree of growth anyway, measures such as conjunctive management may be needed to avoid what has always happened in the past— namely, that growth in a zero-sum context shifts water to human uses at the expense of environmental needs. To assume that keeping the water pie fixed will somehow lead to a cessation of growth and a restoration of the environment seems to us unrealistic in light of the past history and present situation in the West. Our interest in offering recommendations to facilitate conjunctive management therefore grows from our recognition that conjunctive management can be used to recognize and realize environmental values, just as it can be used to recognize the more traditional values of development.

Relating Other Water-Policy Proposals to Conjunctive Management

Western states, including the three we have studied, face common challenges to improve water management generally and conjunctive management specifically. The policy recommendations we examine here are common among water-policy scholars. In linking these recommendations to conjunctive management, we underscore their importance and draw attention to their practical implications.

More Completely Specified Water Rights

More complete specification of property rights, particularly in groundwater, has been a frequent policy recommendation. As long as water rights are unquantified, uncertain, or tied to particular uses or locations, it is more difficult to resolve conflicts over scarcity and avoid the costs created by those conflicts.

Recently, improved specification of surface and groundwater rights has been urged in relation to stream adjudications that are occurring in a number of western states, sometimes in an effort to settle long dormant Indian and federal reserved rights claims. If the effects of groundwater pumping on stream flows are not legally recognized, a significant institutional barrier lies in the path of resolving the scarcity issues at the heart of any such adjudication. Better specification of groundwater rights, and matching those groundwater rights systems with the rights to use surface water, would ease the task of addressing and controlling the effects of groundwater pumping on surface flows.

More complete specification of property rights has been advocated and used also to limit groundwater overdrafting. Where groundwater rights are indeterminate or subject to multiple competing legal theories, as in California, groundwater users have little incentive to cooperate in restraining pumping or to coordinate basin recovery efforts. Basin adjudications in California, and the creation and allocation of groundwater rights in Arizona, helped slow or cease overdrafting by limiting access and pumping. California's approach of securing groundwater rights through basin-specific adjudications, however, has surely been a more costly and time-consuming means of achieving that result than Arizona's adoption of a groundwater rights system as a matter of state policy.

Better specification of water rights, especially groundwater rights, would also facilitate water transfers. Economists promote transfers as a means of moving scarce supplies from lower-valued to higher-valued uses, and this is undoubtedly true, but that language may not convey some of the other practical ramifications of transfers. Transfers are also related to the promotion of water conservation and water-use efficiency, because water users have little incentive to restrain water use or invest in efficient devices if they cannot capture any benefit from their water savings. Greater specification of water rights is linked not only to the prospects for transfers themselves, but also through them to realizing the potential for water conservation in the West.[1]

Whether motivated by a desire to protect groundwater basins, ease stream adjudications, facilitate water transfers, or promote conservation, better specification of water rights is likely to encourage conjunctive management. Water users are more likely to participate in conjunctive management if they can control and actively manage surface water and groundwater, and if they can rely on having access to a specific share of the yield of a water resource. Conjunctive management programs often involve water users exercising restraint with respect to one water source—reducing pumping, for example—while meeting their needs from another; assurance that one's future rights of access to a resource are not compromised by current restraint may be essential to putting resource use on a more sustainable path. Rights to store and recapture water, as exist in Arizona and in some of the adjudicated basins in California, further expand water users' willingness to participate in conjunctive management. Thus, in addition to the other arguments that have been made on behalf of improved specification of water rights, there is a case to be made for it as a means of promoting conjunctive management.

The form or structure of water rights will affect the purposes and practices of conjunctive management. As discussed in previous chapters, the manner in which water rights are specified may result on the one hand in conjunctive management to maintain water table levels that support stream flows, as in Colorado, or on the other hand in conjunctive management that takes advantage of the storage capabilities of groundwater basins, as in Arizona and California. As states and localities improve the specification of water rights, but do so in diverse ways, we are likely to see conjunctive management become more widespread but in a variety of forms.

Regional and Interstate Water Banking

Water banking—collectively organized arrangements for storing and/or exchanging surplus water or unexercised rights—has also received substantial support in the water-policy literature. One important argument for water banks has been to support water transfers. By acquiring water rights or storing surplus supplies and making them available for others to acquire and use, state or regional authorities can reduce the search, bargaining, and negotiation costs among buyers and sellers. In addition to promoting water transfers, water banking can support conjunctive management. Water banks ease the movement of water across sources and among users and uses—"banked" water may be surface or underground, from agricultural or urban users. The surplus rights or supplies made available through water banks can be used in conjunctive management projects—indeed, where physical supplies are being banked rather than just paper rights, water banks often develop or participate in conjunctive management projects to store water until buyers are ready to claim and use it.

Regional and even interstate water banks are, in fact, starting to take shape. California's Drought Water Bank in 1991 and 1992 has been regarded as a successful response to increasingly difficult conditions during the final two years of the 1987–1992 drought, and the state clearly plans to return to water banking as needed in the future (California Department of Water Resources 1993). In Colorado, water banking efforts in the Arkansas basin are ramping up. The goal of those efforts is to allow unused stored project water to be made available to other users rather than spilled from reservoirs, thus providing an additional source of water for conjunctive management. The agreement between Nevada and Arizona discussed in Chapter 8 may be the first interstate banking arrangement in the Colorado River Basin, but it is unlikely to be the last.

Land-Use Regulations to Protect Water Resources

At least three policy recommendations for integrating water resource considerations into land-use planning and regulation have important implications for conjunctive management. As with better specification of property rights and the use of water banks, these policy proposals may be introduced for other reasons or serve other purposes but nevertheless benefit conjunctive management.

Protection of Recharge Zones. The role conjunctive management will play in the future of the West's water resources remains to be seen, of course, but that role is likely to be greatly diminished or foreclosed altogether in some locations if land-use regulations do not protect areas of groundwater–surface water interface (U.S. Advisory Commission on Intergovernmental Relations 1991). If development and urban sprawl continue without recognition of those interface sites, communities hoping to turn to conjunctive management at some time in the future may find that they have paved over and built on the very sites best suited to groundwater recharge.

Regulations that protect the groundwater–surface water interface may have been designed to meet other goals. Protection of wetlands, stream channels, and riparian habitat may be seen as primarily serving environmental needs— or criticized as placing the protection of other species above the interests of people—but it has a practical value for human communities as well. In addition to preserving biodiversity and areas valued for their aesthetic or recreational amenities, such regulations are also likely to preserve the sites where surface water flows can move to underground storage, or where groundwater supports the base flow of surface streams. Those locations have enormous potential value for the development and implementation of conjunctive management programs.

Prevention of Groundwater Contamination. Regulation of the storage and disposal of potential toxic or pathogenic materials above ground or in underground storage facilities is appropriately designed for public health protection from groundwater contamination, but such regulations also have vital benefits in preserving opportunities for conjunctive management. Conjunctive management entails viewing a groundwater basin not merely as a source of water supply but also as a water storage reservoir. Measures that control overlying land uses in ways that guard against contamination dangers, and/or punish individuals who allow contamination incidents to occur help protect a basin's water-supply yield and its value as a storage reservoir. Protected basins have greater long-range potential to be part of a conjunctive management program than basins in which water quality is vulnerable or already has been compromised.

Assured Water-Supply Requirements. As discussed in Chapter 8, these requirements have been motivated in many instances by desires to restrain development. They have the additional effects, however, of drawing the attention of developers and local governments to the issue of water-supply reliability and measures that may enhance it. Where conjunctive management is physically feasible, its development and implementation may be encouraged by regulations that require assurances of water-supply reliability as a condition of development approval.

Financial Tools

Conjunctive management projects, like many other water management improvements, require initial and ongoing financial resources. It is unrealistic to expect that such commitments will come from state and local government general-fund budgets, which are lean even in good economic times and are devoted to other high-priority governmental services and commitments. It is probably also unrealistic over the long term to rely on water users' membership in voluntary associations as a sole means of supporting conjunctive management operations, despite the relative success of groups such as the Groundwater Appropriators of the South Platte (GASP) in Colorado, as water users who are benefited by conjunctive management retain the option of not participating financially.

Sustainable financial support for conjunctive management projects can be crafted to come from water users rather than general-fund budgets, and in ways that enhance the project rather than merely fund it. "Pump taxes," which are fees assessed on the basis of the amount of water used, have the virtue not only of funding a groundwater management program but also adding an incentive to restrain pumping. Flexible pump taxes, like the ones used in a couple of California groundwater basins, can be raised or lowered from year to year based on groundwater and surface water conditions, further facilitating conjunctive use. An alternative idea recommended by the Governor's Groundwater Commission in Arizona in 2001 was a "mined" groundwater tax, revenues from which would support coordinated planning among water users at the Active Management Area (AMA) level and fund the replacement of mined groundwater through recharge operations. Fees may also be collected from entities that store water in a basin—"rent," as it were—and those revenues can also support the conjunctive management program that makes the storage possible. By any of these ideas or combinations of them, states can empower water districts or other local or regional governments to collect revenues that help support the development of sustainable and efficient conjunctive management projects.

State-Specific Policy Recommendations

The ideas discussed in the preceding relate to any state. We also have some specific recommendations for California, Arizona, and Colorado, taking into account the circumstances and potential of each state.

California

- Choose among or combine the competing legal bases of groundwater rights, and select a system in which groundwater rights would be specific and transferable.
- Clarify individuals' rights and basin agencies' authority to store water underground and recapture it later.
- Provide the CALFED program with adequate state and federal funding to develop and implement offstream storage projects as part of the final plan for the Bay–Delta situation.
- Revise the Department of Health Services regulations concerning groundwater recharge with reclaimed water, to aid Californians in achieving greater water self-sufficiency while protecting the health of water consumers.
- Establish and maintain a statewide database or directory of conjunctive management projects.

Uncertainty about groundwater rights in California is a substantial and persistent impediment to more effective groundwater management generally, and conjunctive management in particular. With respect to groundwater withdrawals, underground water storage as part of an effective conjunctive management requires two kinds of specificity. One is a collective specificity of the total amount of groundwater rights that can be exercised in a basin, which is essential in determining how much replenishment is needed not only to meet pumpers' demands but also actually to add water to storage in the basin and be reasonably certain of its fate. The other is the individual-level specificity that assures groundwater users who restrain their pumping so as to participate in a replenishment program, especially in-lieu substitution programs, that they are not merely forfeiting groundwater to other users. Currently, the California Water Code contains language stating that temporary restraint in the exercise of water rights shall not constitute a forfeiture, but when groundwater users are unsure what rights they have in the first place such language does not provide the assurance it would otherwise. Except in the adjudicated basins, these specificities do not exist in California, and as described in Chapter 4 the prevalence of multiple competing theories of groundwater rights in the state provides nothing but an atmosphere of confounding uncertainty that thwarts efforts to control pumping, reduce overdraft, promote conservation, facilitate transfers or, to say the least, support conjunctive management.

Judges or legislators who attempt to construct a unified doctrine of quantifiable groundwater rights in California will surely pay a political price, but the benefits of doing so could resound statewide for decades to come. Adoption and implementation of this recommendation likely will require a coalition of water agencies, business organizations, and the environmental community to increase the pressure on state officials until the current deadlock finally ends and needed institutional reforms occur.

Better specified rights to groundwater withdrawals will advance conjunctive management in the state only so far, however. California policymakers would do well also to adapt a regulation similar to the Arizona statute that protects the right of an individual or organization to recover a quantity of water it has stored underground. Because California does not have such a state-centered program of water management as does Arizona, a legally protected storage-and-recapture right could be chaotic in the absence of an authority to monitor and account for stored water. Basin management agencies in California, which as we have seen may be overlying special districts, court-appointed watermasters, or the agencies responsible for implementing Assembly Bill (AB) 3030 plans, should also be authorized to establish water storage accounts and provide periodic reporting of stored water quantities, in addition to monitoring groundwater levels to protect the interests of overlying landowners and other basin water users. Such a change can reduce the transaction costs groundwater users face in developing water storage programs, leading toward a future in which conjunctive management projects could be initiated more easily by any individual or organization with rights in a basin.

The CALFED program has been controversial since its inception, and it may seem like folly to include any policy recommendations concerning it. Our hesitation to do so is outweighed, however, by our concern about what might come out of the CALFED process if funding for offstream storage is not included. Without provisions for enhancing water storage in wetter periods, the

CALFED process will face great difficulty in reaching any consensus recommendations for reoperating the facilities of the State Water Project to preserve aquatic species and habitat in the Delta region, and in implementing the reoperation whether or not it was agreed by consensus. Especially in light of the possible effects of climate change mentioned in Chapter 8, but even if those effects do not occur, additional storage capacity upstream of the Delta will be needed to divert and capture heavy springtime river flows for controlled releases throughout the year to manage salinity levels in Delta streams. Additional storage capacity south of the Delta, to which some surplus flows may be directed in wet years, will allow the reoperation of the State Water Project to occur with smaller impacts on central and southern California residents and farms. That protection is key to obtaining the political support needed for CALFED success, and could also enhance the prospects for conjunctive management in the drier southern half of the state. Some stored surplus water taken from the Sacramento–San Joaquin River system during especially wet periods could be stored underground in overdrafted Central Valley aquifers for use in drier years when a higher proportion of river flows must be dedicated to environmental needs.

Over the long term, sustainable water use in California means relying more on the resources within the state, and maximizing the reuse of supplies before they are disposed to the ocean. At present in most of the state, treated wastewater is discharged to streams and rivers, and represents a large portion of the year-round flow in Southern California streams. Californians also employ reclaimed water for landscape irrigation to a commendable and increasing degree. As larger quantities of water are treated to meet rising quality standards, however, the need will increase to integrate that growing source of supply into the water-delivery system for a wide variety of purposes.

Purified wastewater has some real advantages as a source of groundwater replenishment water. First, it is a relatively steady supply, which could make for more economical use of recharge facilities throughout the year. Second, the quality of purified wastewater is typically better than that of the stream flows that are currently captured and diverted into recharge facilities.[2] Currently, however, California Department of Health Services regulations restrict the prospects for groundwater recharge with purified wastewater. With rigorous monitoring regimens and the nation's strongest water quality standards, the department could fulfill its obligation to protect public health while allowing more replenishment of aquifers with treated water. To do so would advance the options for conjunctive management and promote a more sustainable and self-sufficient water future for the state.

Finally, if California is to continue its tradition of locally initiated and locally managed conjunctive management projects, as we believe it will, the state government could nevertheless play a more supportive role by lowering the information costs local water managers face in learning about conjunctive management practices. Relying on local initiative for groundwater management generally, and conjunctive management in particular, has benefited California by encouraging a variety of conjunctive management methods that are employed throughout the state to match local conditions. Still, water users elsewhere in the state who may be interested in pursuing conjunctive management face quite a task in trying to learn about and learn from the experiences of their counterparts around the state. The only published compendium of conjunctive management projects in the state was produced two years ago by a nongovernmental group, the Association of Ground Water Agencies (2000), which was very well prepared but nonetheless omitted some cases. The task of local water management agencies and organizations in learning about the methods, scale, costs, and performance of conjunctive management in California could be lowered substantially if the California Department of Water Resources, or another appropriate agency selected by the governor or the legislature, would compile a statewide inventory of conjunctive management projects and practices, and update it periodically. The Conjunctive Water Management Branch in the Division of Planning and Local Assistance in the Department of Water Resources has taken a step in this direction, but mainly with respect to the newer initiatives that have received financial support from recent state water bond issues rather than the other conjunctive management programs that have operated around the state for decades. Such an information compilation and dissemination effort would allow the state to play a supportive role without altering its long-standing practice of leaving the development and implementation of conjunctive management projects to local agencies.

Arizona

- Revise statutes on special districts to allow multiple jurisdictions greater flexibility for coordinated fund-raising for the infrastructure needed for conjunctive management.
- Offer more incentives and opportunities to rural communities and areas outside of the AMAs for devising conjunctive management programs.
- Devise clearer rules for sub-basin management.

The first two recommendations follow from our discussion of problems facing Arizonans that may hinder the future of conjunctive management: the inability of local water providers to coordinate infrastructure financing and the lack of assurances for storage and recovery of water supplies needed for conjunctive management programs outside of AMAs. For many small-scale water providers, these factors can significantly raise the costs of engaging in conjunctive management. Conjunctive management requires access to a reliable source of surface water or treated effluent for storage. As noted in Chapter 5, without opportunities to coordinate the financing of surface water infrastructure or effluent treatment, small-scale providers may not have the capacity to develop capital-intensive projects. Moreover, outside of AMAs, water providers with surface water or effluent available for storage cannot engage in long-term underground storage without risking the loss of that water to groundwater users in that basin.

The third recommendation addresses the potential negative effects of Arizona's existing recharge regulations and its AMA-centered approach to groundwater management. For conjunctive management programs to be sustainable well into the future, water providers need to pay attention to the overall appropriateness of storage and recovery sites in terms of water quality and environmental impacts. Because each AMA is governed as a single, uniform resource—a single basin, as it were—water recharged anywhere in the AMA yields storage credits that can be used to extract water anywhere else in the AMA. Accordingly, little attention has been given to the potential negative or positive impacts of conjunctive management on hydrologically sensitive areas within AMAs. We encourage Arizonans to identify these areas and establish stricter rules for conjunctive management programs to avoid excessive pumping in sensitive areas without sufficient recharge in the same sub-basin.[3] At the very least, conjunctive management programs that were intended to redress overdraft problems should not be allowed to make them worse because institutional rules fail to relate recharge to pumping.

Recent efforts to consider reforms to the Arizona Groundwater Management Act have recognized the need for some of these types of policy changes. One of the recommendations of the Governor's Groundwater Commission, in fact, was to devise enabling legislation for an AMA Infrastructure Financing authority to aid some water providers in cooperatively financing the infrastructure necessary for utilizing renewable water supplies. Another recommendation was to locate and study more thoroughly sensitive riparian areas.

These recommendations point to a broader issue Arizonans may need to grapple with: each region and locality has divergent environmental and economic conditions that may demand more locally directed efforts to achieve efficient water management. Arizona's centralized system of water governance does not necessarily facilitate locally driven efforts to address local problems.[4] Although Arizona is not a home-rule state like California and Colorado, there is no reason why the state legislature could not authorize more locally directed groundwater management efforts.

Colorado

- Treat tributary ground and surface water supplies as one source, not just in law but also in practice.

Conjunctive management can serve numerous purposes—support stream flows, store surplus surface water when available for use at a later date, or maximize the amount of water available for use at any given time by requiring water users to utilize surface water when abundant. The first purpose is the most restrictive and limited use of conjunctive management. The third purpose is the least restrictive and takes advantage of all of the benefits of conjunctive management.

Currently, for institutionally driven reasons, Colorado water users overwhelmingly use conjunctive management for the first purpose, to satisfy interstate compacts and to protect investments based on the seniority system. There are shortcomings of using conjunctive management exclusively to support stream flows, however. First, vast amounts of groundwater are made inaccessible because withdrawals would affect surface water rights holders. The South Platte Basin is estimated to hold 8 million acre-feet of water. Significant portions may be unusable because of depth and water-quality issues, but a considerable amount of groundwater remains to be tapped and actively managed.

Second, under the current system the amount of groundwater pumping is entirely a function of the availability of surplus surface water. Surplus surface water is diminishing in both the South Platte and Arkansas River basins as municipalities make more extensive use of their project water. In addition, pressing concerns over water quality and endangered species mean that water once available from the South Platte from late fall to early spring is likely to be kept in the stream and not appropriated for off-channel uses. Most tributary groundwater pumping rests on a foundation that is at best temporary and at worst disappearing—the supply of surplus surface water. At some point in the not too distant future Colorado will be faced with an inevitable choice—severely curtail tributary groundwater pumping though millions of acre-feet of usable groundwater are available, or find different means of satisfying senior rights, interstate compacts, and endangered species demands. In other words, Colorado will have to choose between forbidding most tributary groundwater pumping and finessing existing institutional arrangements to allow more active use of tributary groundwater basins. We believe that Colorado will choose the latter course of action.

In practice, what would that mean? It would mean treating tributary groundwater and surface water as a single source in practice and not only in law. Senior rights would be satisfied with the type of water—ground or surface—in greatest abundance. During relatively wet years, senior rights would be satisfied much as they currently are, through the diversion of surface flows. During drier years, senior rights would be satisfied through tributary groundwater pumping. In addition, more active and widespread groundwater replenishment programs would have to be instituted to ensure that basins are adequately recharged during wetter years. Currently, only the most senior water rights are satisfied during dry years even though substantial amounts of tributary groundwater are available in the basin. Under this new system, tributary groundwater would be used during dry years to buffer most water users from drought.

More active management and use of tributary groundwater is institutionally feasible. As noted in Chapter 6, existing institutional arrangements and water organizations already coordinate water allocation and use among users of surface and tributary groundwater in the South Platte and Arkansas basins. Increasing the active use of groundwater basins would require such organizations to coordinate their members' water use more closely, and, in conjunction with division engineers and water commissioners, coordinate activities to ensure adequate stream flows for interstate compact and endangered species commitments. Water organizations, well-owners' associations, conservancy districts, conservation districts, and irrigation districts would be allowed to pool ground and surface water rights of their members, satisfying those rights with the most abundant type of water. Such organizations would also be required to engage in groundwater replenishment activities to restore water pumped during drought years. Water commissioners, water courts, and division engineers would continue to work to protect and enforce water rights and to settle conflicts among water users.

In the early 1970s, Colorado flirted with closely integrating surface and groundwater use when augmentation plans were first adopted in the South Platte Basin. GASP, for instance, was allowed to pump tributary groundwater to satisfy senior surface water rights. The state engineer, however, backed away from such practices as senior rights holders who were not members of GASP expressed concern over the security of their rights. It may now, three decades later, be an appropriate time to begin experimenting with such practices once again. The state engineer has invested heavily in state-of-the-art decision support systems allowing for more accurate and timely tracking of water availability and use. Water users and water administrators have extensive experience with augmentation, recharge, and replacement plans. In addition, the legislative and judicial decisions of 2002 have precipitated changes in the processes by which augmentation plans are reviewed, which may open a window for a broader reconsideration of how to reconcile groundwater pumping and surface water use.

Colorado's increasing urban and environmental water demands can be met only by eliminating agriculture, which is politically and economically infeasible; building additional large surface water storage and distribution systems on the western slopes of the Rockies, which is expensive and complicated by changing conditions in the Colorado River basin; or more fully incorporating and actively using tributary groundwater within existing institutional arrangements. We believe that the more active use of tributary basins is the prudent option.

Conclusion

The potential of conjunctive management to address the West's pressing water supply and demand dilemmas now and in the future rests largely on the continued efforts of citizens, water managers, and policymakers to encourage institutional change. Skillful promotion of institutional change within any state requires an understanding of how current institutional arrangements are connected with conjunctive management practices and other aspects of water-resource management. Those connections are themselves shaped by other aspects of the state's historical development and political culture.

In this book, we have tried to convey insights into the connections between institutions and conjunctive management, particularly how different institutional settings shape water users' incentives to participate in or avoid conjunctive management. In doing so, our effort has been to move beyond simplistic explanations that appear in the water-management literature: institutions matter, or institutions are the problem, or there exists one clear path for improving institutional outcomes. Achieving this meant conducting a comparative study of actual institutional settings and conjunctive management practices—in our case through a study of Arizona, California, and Colorado.

In those states, with the help of dozens of individuals and organizations, we delved inside the institutional black box to identify the key features of water-rights institutions and water-management organizations that shape conjunctive management. In many instances institutions impede conjunctive management, but in others they provide the key to user restraint, water storage and recovery, and interorganizational coordination, all of which are needed for conjunctive management. The interactions between institutional arrangements and water management choices are complex, and simplified institutional solutions are not likely to provide many answers to the West's water supply dilemmas. Instead, the policy options that are suited to conjunctive management, or any other technique of improved water-resource management, must grow from a comparative understanding but be applicable to a particular setting.

The policy recommendations we have presented in this chapter are therefore not entirely new and groundbreaking. Our goal has been to contribute a stronger understanding of how institutional changes relate to the incentives to engage in more efficient water use. In western water management, institutional change will require considerable debate and struggle before citizens and water users can agree on new rules and how they will be implemented. Debate and struggle are not indicators of the policymaking process going awry—when the stakes are high and rising, as they are, vigorous debate and the struggle to reach supportable decisions are the very essence of policymaking at work. We hope this examination of institutions and conjunctive water management has provided some new insights to inform those discussions and the decisions that lie ahead.

Appendix

The Three States: Why We Chose Them, and What We Did

As noted in the Preface, the data on conjunctive management activities and institutions described in this book were gathered under a study conducted between 1997 and 2000, funded by the National Science Foundation and U.S. Environmental Protection Agency. Here we describe the research methods used for data collection and analysis.

We collected the data by devising a set of coding forms that consisted of structured questions on the type, purpose, and performance of conjunctive water management projects, the organizations that manage them, and the laws and policies governing ground and surface water uses in Arizona, California, and Colorado. The coding forms were completed through interviews we conducted between 1997 and 1999 with water providers and water management organizations involved in conjunctive water management projects. Some secondary sources were also used to complete the coding forms.[1]

Research Framework and Study Design

The design of our study was based on a larger body of work in institutional analysis that has been developed over the past 30 years at Indiana University, most notably by Elinor and Vincent Ostrom and colleagues at the Workshop in Political Theory and Policy Analysis (see Ostrom 1990,1999). Institutional analysis relies on in-depth comparisons of the structure and performance of institutional arrangements to explore the effects of institutional arrangements on human choices and actions. This research paradigm recognizes that institutional arrangements are processes of intentional human design, constrained by physical and community circumstances. It provides a systematic framework for identifying key variables and levels of analysis that are common to most institutional settings.[2]

In using the institutional analysis framework, our research methodology emphasizes comparisons of conjunctive management activities on a variety of dimensions. First, we inventoried multiple forms of institutional arrangements. This inventory consists of public and private organizations involved in the development and implementation of conjunctive management programs, including regulatory and oversight jurisdictions. We collected data on the operating procedures and regulations of the organizations involved in conjunctive management activities. We also completed forms on state laws and regulations that define rights of private and public entities to capture, store, retrieve, extract, and transfer water supplies and researched relevant federal, state, and local environmental protection or land-use regulations that may affect conjunctive management programs.

Community context and physical characteristics of field settings also are important factors in an institutional analysis approach. Therefore, we designed the coding forms to include relevant information on the hydrological characteristics of conjunctive water management sites, as well as factors such as the size of the community of water users where conjunctive management projects occur and the communication patterns among water providers in these locations. In addition, we included contextual factors describing the organizations' operating projects, such as the purpose or type of these organizations, together with their membership or size and financial assets.

The third aspect of our research design focused on the performance of conjunctive management projects. To evaluate fully the successes and failures or effects of institutions, the institutional analysis framework emphasizes the relative performance of institutions in solving the dilemmas that these institutions are designed to address. In our study, therefore, we designed our coding forms to identify the goals of conjunctive water management projects and the impacts of those projects over time. We included performance measures such as the amount of water withdrawn and stored by water users, the quantity of water rights exchanged, the costs of pumping and diverting water, the quality of water stored, the costs of project production, and the transaction costs associated with projects.

Identifying Conjunctive Water Management Projects

Before collecting data, we identified conjunctive management programs across the three states. The process of collecting data began with the identification of populations of water organizations conducting conjunctive management in each state during 1997, or for any of the 10 years prior to 1997. We chose 1997 as a base year to compare the performance measures of these projects.

In Arizona, identifying the population of active projects in 1997 was relatively straightforward. The Arizona Department of Water Resources issues permits for all organizations actively recharging groundwater or taking surface water "in-lieu" of groundwater. The State also issues permits to organizations for the recovery of any surface water that has been stored underground for future use. Arizona, therefore, has standardized records of the complete population of projects and of all the organizations involved in those projects. Because the population of active projects in 1997 is relatively small (<50, the study included the entire population of organizations and jurisdictions permitted to run conjunctive-use projects.

Arizona Department of Water Resources records indicate that 42 conjunctive management projects held permits in 1997/1998. Although they possessed permits, some of these projects were not actively engaged in conjunctive management activities during 1997. In the Phoenix area, three projects had become inactive since the mid-1990s and had only been in operation for a couple years. Five of the thirty permitted projects in Phoenix had not begun to operate because they were new or experienced technical problems. Four permitted projects in the Phoenix area were pilot projects and had reported limited activity to the state. In the Tucson area, one of the projects was not yet operating in 1997, while another had just become active in 1997. Therefore, only 33 of the permitted projects in Arizona were actively engaged in conjunctive water management activities in 1997, but 6 more new projects were still involved in the planning and organization of conjunctive management activities. The three projects in the Phoenix area that were no longer active in 1997 were included in the population because they had engaged in activity in the past 10 years and maintained the permits and facilities to reestablish activities in the future. Of the 42 permitted projects, only 5 had been in operation prior to 1992. Thus, most projects that we observed in Arizona are relatively new.

In California and Colorado, the population parameters of conjunctive management projects and data accessibility differ widely from the Arizona projects. California has no centralized source of information on conjunctive management operations because groundwater is managed locally. It was not feasible to identify all of the organizations involved in conjunctive management by contacting California's water providers because there are thousands of water providers in the state.

To identify the organizations engaged in conjunctive management, we used a 30 percent cluster sample of California's 450 groundwater basins, which the California Department of Water Resources groups into seven hydrologic regions. Before sampling the basins, two hydrologic regions were eliminated as unlikely to contain any conjunctive management activities based on their physical conditions and low water demands.[3] Sampling from the remaining five hydrologic regions yielded a sample of 70 basins.

With that sample, we used a California Department of Water Resources list of water agencies by county and zip code to identify water providers operating, or potentially operating, in the sample basins. We then contacted water providers by telephone to determine the type of water management activities and groundwater governance institutions operating in the sample basins, and specifically whether any conjunctive management activities occurred there. Twenty-three sample basins turned out to be in remote locations with little developed water use and no organized water-resource management. In the 47 developed basins in the sample, we identified conjunctive management activities underway in 12 basins and planned programs in 4 others.

As in California, the State of Colorado does not systematically track conjunctive management of surface and ground water. The State of Colorado administers water laws and regulations through seven divisions of the state engineer's office, which are organized around the state's major river basins. Each of the state engineer divisions is further divided into multiple districts. Each division engineer's office maintains hundreds of records of augmentation plans and recharge diversions, which involve the conjunctive use of ground and surface water. These records do not specify the type of augmentation taking place, or whether it involves conjunctive water use. Therefore, gathering data on conjunctive management activities required contacting water providers individually. As in California, contacting all of the water providers in Colorado was not feasible, so a sample of Colorado's water districts was taken.

To identify projects in Colorado, the study used a 30 percent cluster sample of the 59 water districts included in five of the state's seven watershed divisions. Two divisions were not included in the sampling because information from the state engineer's office indicated that no conjunctive water management was occurring in them. The final sample totals 17 districts, including five from the South Platte River division, four from the Arkansas River, three each from the Colorado and Gunnison River divisions, and two from the Rio Grande River division. Records from the division engineer's offices provided information on the augmentation plan filings in each sample district and identified water providers involved in these plans. Within these sample districts, we identified 42 conjunctive management projects.

For all three states, we relied on water management organizations that supply or produce conjunctive management projects as the source of primary information for completing coding forms. Organizations that provide or fund these types of water management services include municipal water departments, irrigation districts, private water companies, and special groundwater management districts. We also examined organizations that participate as providers or producers of components of conjunctive management services. These organizations often include federal or state water projects that supply surface water, state water banking authorities, or "wholesale" water management districts that contract for surface supplies on behalf of member organizations in the district.

Through our interviews with organizations involved in conjunctive management projects, we collected data on the purpose of projects, the costs of projects, the impacts or outcomes of these projects, and the types of organizational arrangements involved in producing conjunctive water management. For some of the data on conjunctive management projects, the organizations that operate them, and the physical conditions of project locations, we also relied on secondary data sources to supplement primary sources. These sources included reports from the Arizona Department of Water Resources, the Colorado division engineers' offices, the California Department of Water Resources, and additional reports by local water agencies. Engineering and hydrology studies by water agencies and state departments of water resources also provided data on the physical conditions of groundwater basins and surface water flows in each state. Federal and state sources provided macroeconomic and demographic data on the three states and their counties.

Finally, we utilized secondary sources of data to identify state and local laws and policies that influence conjunctive water management activities in each state. Archived resources yielded data on state constitutional and statutory provisions related to water, as well as court decisions. Unlike the project and organization forms, data collection on state laws was not conducted during the interview process. To complete these forms, we researched water statutes and constitutional provisions using *LexisNexis* and additional legal reference sources. Rules and regulations established by the Arizona Department of Water Resources and the Colorado Division Engineers for the administration of water rights provided additional data sources. In California, where state statutes or court adjudications have created the authority for local regulation of groundwater resources, data on local-level rules also applied to the projects found in the sample. This research process provided a complete set of data on the rules governing conjunctive management.

Evaluating the Data

We emphasize descriptive data on each state's institutional setting, including state groundwater and surface water doctrines and the organizations involved in water resource management. We then consider descriptive data on the types of projects of interest, the types of organizations that operate them, the location of projects, and the impact of these projects in the state. Based on this evidence, we suggest how each state's institutional setting has facilitated conjunctive water management projects. From there, we evaluate the comparative differences across the three states to draw conclusions about the role of institutions in conjunctive water management.

The coding forms provided a rich set of data from which we conducted different descriptive and quantitative analysis. In evaluating the relationship between institutions and conjunctive water management, comparisons were made at the state level, sub-state level, and project-level. The multiple levels of analysis leverage the data set by providing more observations and more measures for drawing empirical inferences compared to using one of these levels alone (King et al. 1994). Specifically, we considered the role of institutions at the substate, or basin-level in California, which provided four different forms of groundwater governance variables across 70 cases. In looking at state-level institutions, the data set allowed only for an analysis of the various forms of water governance institutions across three cases. Despite the limited number of state-level cases to compare the role of institutions, the cross-state comparison was able to examine multiple measures of conjunctive management projects to compare the different effects of institutions. These included conjunctive management project locations, institutional boundaries, project types, organizations involved in projects, and the institutional form of those organizations.

Notes

Chapter 1. Water Scarcity, Management, and Institutions

1. The summer monsoon storms in Arizona also contribute a large portion of that state's scant average annual rainfall.
2. Of course, if overdrafting is too severe, this can lead to soil compaction, which reduces the overall storage capacity of a groundwater basin.
3. California, for example, illustrates the combined effects of deliberately practiced conjunctive management and this incidental conjunctive use by water users with access to multiple water sources—in wet years, groundwater provides only 30% of water used in the state; in drought years that percentage has risen to 60% (Association of Ground Water Agencies 2000, 4).
4. To illustrate the potential of conjunctive management in the new era: Southern California's newest and largest surface water reservoir—Diamond Valley Lake, constructed and owned by the Metropolitan Water District of Southern California—cost $2.5 billion and took 10 years to complete. The estimated storage capacity of the groundwater basins in Southern California—21.5 million acre-feet—is enough to fill that surface reservoir 26 times (Association of Ground Water Agencies 2000, 2). Little wonder, then, that the Metropolitan Water District's Integrated Resources Plan calls for a doubling of groundwater yield from conjunctive management projects—from 100,000 acre-feet per year in 2000 to 200,000 in 2020 (Metropolitan Water District of Southern California 1996a).
5. We follow the distinction made by Kiser and Ostrom (1982) and North (1990) between institutions and organizations. Institutions are rules, norms, and strategies. Organizations are "groups of individuals bound by some common purpose to achieve objectives" (North 1990, 5). For instance, organizations include legislatures, government agencies, firms, and family farms to name a few. For applications of institutional analysis to water and other natural resource issues, see, for example, Young (1982), Ostrom (1990), Blomquist (1992), Tang (1992), and Heikkila (2001).
6. This approach was supported recently in the National Research Council's report, *Envisioning the Agenda for Water Resources Research in the Twenty-First Century.* Recognizing the need for research that could aid in the development of improved ground-

water management, particularly conjunctive use, the council's report recommends comparative studies of water law and institutions, and of institutions that have been organized by water users (National Research Council 2001). The council acknowledged that "research on institutions occupies only a very small portion of the current water research agenda," (*9*) which may be a reason why "the nation has accumulated over a century of experience with a variety of water policies and management modes, yet we have not learned as much as we might from that experience"(*40*). Accordingly, the report recommends "that efforts should be made to invest relatively more in institutional research than has been the case in the past" (*33*).

7. Others have also recognized these differences and their relevance to conjunctive management—for example, "Western states vary considerably in whether and how they attempt to achieve conjunctive, effectively and efficiently coordinated, management of surface and ground water. Some states, such as Arizona and California, rely primarily upon water supply districts or other organizations. Others, such as Colorado and New Mexico, attempt this coordination within the framework of water rights law" (Gregg et al. 1991, *103*).

8. These figures represent withdrawals for water use and not necessarily *consumptive* use. When water is withdrawn for use, some of that water may later be returned to rivers or streams. The portion of water that does not return to the stream is considered consumptive use. Consumptive water use in agriculture represents a larger portion of withdrawals than most municipal or industrial water uses (Reisner and Bates 1990).

9. Most western states maintain separate doctrines for governing ground and surface water supplies, yet many of these states have struggled with the interrelated nature of ground and surface water supplies. See Glennon and Maddock (1994) for an analysis of the inconsistencies in Arizona case law in recognizing the relationship between groundwater and surface water.

10. For more thorough descriptions of the state–local relationships in the three states, see Krane et al. (2001).

11. This observation is not only one of the most interesting insights of our work, but it also validates and underscores the value of comparative analysis. Had we focused our study of institutions and conjunctive management on one location or state, we could easily have missed this point.

Chapter 2. The Promise of Conjunctive Water Management

1. In this book, we refer to both groundwater basins and aquifers. Although in many contexts the distinction will not matter, where it does the distinction for our purposes amounts to this: a groundwater basin may include more than one aquifer, but not vice versa. A groundwater basin may be thought of as a *geographic* phenomenon, as if from an aerial view one could peer through the earth's surface and delineate the boundaries of an underground drainage area toward which water moves or within which water collects. An aquifer may be thought of as a *geologic* phenomenon—a stratum of permeable, water-bearing material. Aquifers exist in layers in some locations, one stratum of water-bearing material lying above another with rock or clay or other less permeable material sandwiched between and separating them. This layering possibility makes it possible for one groundwater basin to contain multiple aquifers.

2. See National Research Council (1997, *38–39*) for additional examples such as these: "Subsidence can also cause flooding, particularly in coastal areas. Between 1906 and 1987, land in the Houston/Baytown region of Texas subsided by between 1 and 10 feet,

resulting in pronounced flooding of valuable land adjacent to Galveston Bay .... The most dramatic example of subsidence is found in the San Joaquin Valley of California, where land surfaces have fallen up to 40 feet in some areas."

3. The terms "natural" and "artificial" have been used in different ways. Some authors have employed them to distinguish replenishment water sources; capturing precipitation or runoff-related surface water flows from within the same watershed and storing them underground would therefore be "natural recharge," while importing the replenishment water from another basin or recharging an aquifer with treated wastewater would be "artificial recharge." Other authors have used the terms to distinguish the physical processes by which the replenishment water gets to the aquifer, so percolation of water through stream channels or lake beds is regarded as "natural" while moving water underground through constructed percolation ponds or by means of injection wells is "artificial." These are, of course, overlapping expressions—replenishment water could be imported into a watershed (thus, "artificial" by the first definition) and stored underground by means of injection wells (thus "artificial" by the second)—and the overlap significantly impairs their usefulness. For our purposes in this book, we will (1) try to avoid employing the terms at all, and (2) when we fail, use the latter distinction which turns upon the means by which the water gets into the aquifer rather than where the water came from.

4. In Arizona and Colorado where quantified water rights are more typical, in lieu recharge involves the actual exchange of water rights. For instance, an Arizona farmer will purchase rights to a specified amount of federal project water by providing the manager of the federal water project with rights to a specified amount of groundwater. Farmers engage in such water swaps because the exchange rate is favorable to them. In California where quantified groundwater rights are not the norm, groundwater users may be offered incentives to use more surface water and less groundwater. The groundwater saved, however, is not owned by any entity.

5. Building an additional 100,000 acre-feet of storage capacity in order to guard against drought conditions that might occur only once per decade, for instance, produces a substantially different financial analysis than the same quantity of storage if built to supply daily water needs on a year-to-year basis.

6. Glen Canyon Dam can annually generate 1300 megawatts with a full reservoir. As a consequence of the 1992 Grand Canyon Protection Act the Bureau of Reclamation operates the dam to annually generate 500–800 megawatts so as to limit damage to riparian areas in the Grand Canyon (Glen Canyon Institute 2001).

Chapter 3. Opportunities and Obstacles for Conjunctive Management

1. A feasibility study by the Natural Heritage Institute (1998, 9) on conjunctive management opportunities in California's Central Valley concurs with these points, noting: "To estimate the hydrologic potential of the pre-delivery of surface water to groundwater banking in the Central Valley watershed, NHI developed the Conjunctive Use Potential model, or CUP ... based on liberal assumptions about: (1) the existence of infrastructure; (2) a limited scale investment in the direct diversion of high flows to aquifer storage ... ; and (3) the availability of suitable groundwater banking sites."

2. On the other hand, Colorado has taken advantage of the close relationship between ground and surface water to assist in meeting its commitments under the South Platte

River compact. As discussed in greater depth in Chapter 6, surplus surface water is placed in recharge ponds located close to the border with Nebraska. The water percolates into the tributary aquifer and bolsters river flows, helping to ensure that Colorado delivers a volume of water to Nebraska that it committed to provide under the compact.

3. Even if all the necessary physical resources are in place for conjunctive management, it may not be an appropriate tool because of other considerations. In some groundwater basins, water storage and recovery may lead to more water management problems than benefits. Extracting stored water from basins hydrologically connected to surface flows draws down the water table and may reduce available surface flows (Glennon and Maddock 1994; Matthews 1991). Excessive recharge of a basin can move the water table too close to the surface and exacerbate flooding or water-quality problems. Early in the conjunctive management program in Southern California's Chino Basin, for example, basin managers raised the underground water table to the point where the groundwater reached the bottoms of some basements and swimming pools, and mingled with the upper soil layers that contained decades of deposited manure from dairy farms. In some basins, artificial recharge could potentially cause spreading of contamination plumes or aggravate soil compaction in areas susceptible to land subsidence.

    The operational requirements of a conjunctive management project may conflict with other water management efforts. The use of surface water reservoirs for groundwater replenishment in addition to flood control and water supply complicates reservoir-management decisions and increases the chance that satisfying one set of objective will impair another.

4. The conflict among multiple uses is readily apparent in the San Francisco Bay and Delta region, which along with the Central Valley rely heavily on the Sacramento and San Joaquin Rivers. The California Department of Water Resources (CDWR 1998b) calculates the San Francisco Bay region's environmental and instream flow water demands as higher than those of other hydrologic regions in California. The region's environmental water demands were nearly 5.8 million acre-feet in 1995, totaling 80 percent of the region's overall water needs (CDWR 1998b). This water is used to address habitat and water quality problems in the San Francisco Bay/Sacramento–San Joaquin Delta Estuary that have resulted from upstream diversions. The CALFED Bay–Delta program was created in 1995 as a multiagency effort to deal with these problems.

5. Whether allocating water on the basis of seniority is a smart policy decision is a different question; here we are simply observing that the prior appropriation system allocates fairly well specified rights. Consider the contrast between the prior appropriation doctrine commonly used in the western United States and the riparian doctrine used in the eastern part of the country. Under riparian law, water rights belong to owners of land adjoining rivers, streams, and lakes. Riparian owners were required to make reasonable use of the water, but not granted rights in specific amounts of water. Nor could their rights to the use of water be transferred without sale of the adjacent land. For a nice discussion contrasting the doctrines of prior appropriation and riparianism see Rose (1994).

6. A lack of transferable private-property rights to water is often recognized as a barrier to improving watershed use and protection (Bish 1977; Cuzan 1979; Johnson 1992). Privatization, in the forms of full-ownership property rights and open markets, has been advocated for virtually all aspects of surface and groundwater control, including the preservation of instream flows (Huffman 1986), the management of aquifers (Anderson et al.1983), and even aquifer recharge and water storage (Provencher 1993). However, these privatization advocates are often criticized for failing to account for the negative externalities and social costs that private users of water supplies can impose

on other water users, species, and habitats (Bates 1989; Bruggink 1992; DeYoung and Jenkins-Smith 1989: Ingram and Oggins 1992; Metzger 1988; Sax 1994).

Chapter 4. California

1. The California Department of Water Resources estimates that average fresh water supplies available within California, with existing facilities and under 1995 conditions, are 77.9 million acre-feet per year (California Department of Water Resources 1998a,b). If an acre-foot of water is enough to supply five to six persons for a year, California annually receives enough water to support a population many times its current 37 million people. California's water problems arise not from natural conditions of water scarcity, but from other conditions described in this section of the chapter.
2. For example, annual rainfall measured at the Los Angeles Civic Center since 1877 has been as low as 4.85 inches (July 1, 1960–June 30, 1961) and as high as 38.18 inches (July 1, 1883–June 30, 1884). Annual runoff has ranged from an estimated 15 million acre-feet in 1977 to more than 135 million acre-feet in 1983 (Littleworth and Garner 1995, 2).
3. Obviously, desalination of ocean water is possible, and at this time desalination facilities are under active consideration by the Metropolitan Water District of Southern California and some of its member agencies. As a practical matter, however, over the past century the cost of desalination has compared unfavorably with capturing and using fresh water precipitation and runoff.
4. Usable capacity is that which is within economic reach of the land surface—that is, groundwater that could feasibly be pumped and used. The total storage capacity of California's groundwater basins, including the capacity below currently usable depths, is estimated to be 850 million acre-feet (Hauge 1992, 15).
5. See Oshio (1997) and Atwater and Blomquist (2002) for descriptions of the historical and current arrangements for allocating water among MWD member agencies.
6. National Audubon Society v. Superior Court, 33 Cal.3d 419 (1983).
7. County of Inyo v. Los Angeles Department of Water and Power, 32 Cal. App. 3d 795 (1973). The decision in that case did not rest on the theory of public nuisance, however, but on California Environmental Quality Act (CEQA) requirements for an Environmental Impact Report, which Los Angeles was ordered to prepare.
8. City of Los Angeles v. City of San Fernando et al., 123 Cal. Rptr. 1 (1975).
9. The Water Recordation Act (Sections 4999 through 5008 of the California Water Code) established a system of recording groundwater withdrawals in four Southern California counties—Los Angeles, Orange, Riverside, and Ventura. Any entity pumping in excess of 25 acre-feet per year in those counties must report its diversions to the state or risk restrictions against claiming unreported use in a future adjudication of rights.
10. Eighteen groundwater basins in the state have been adjudicated (Association of Ground Water Agencies 2000).
11. The California Department of Water Resources (1998b) has published a table enumerating the number of water supply organizations in the state. The most common types are: county service areas (880), mutual water companies (801), community services districts (309), water utilities (195), county water districts (178), California water districts (157), and irrigation districts (97).
12. AB 3030 does not grant any local government the authority to determine or limit water rights.
13. Baldwin v. County of Tehama (1994).

14. We also want to emphasize that our findings should *not* be misread as stating that conjunctive management is being practiced in only 12 basins in California. We are aware of several other conjunctive management operations in the state, in locations that did not fall into our sample. Descriptions and data on five additional cases, for example, can be found in Blomquist (1992), and we know of at least four others. The Association of Ground Water Agencies (2000) published results of a survey of conjunctive management programs in the state, and concluded that conjunctive management operations statewide yield an average of 2.5 million acre-feet of water per year.

15. Three basins (Antelope Valley, Modesto, and Suisun–Fairfield Valley) appear on both lists. In those basins, water use for irrigated agriculture remains very high but the area is also urbanizing rapidly.

16. This is also referred to as "passive recharge" (Todd and Priestaf 1997).

17. A supplemental observation is appropriate for The Metropolitan Water District of Southern California, as it has facilitated local conjunctive management projects by providing replenishment water at a discounted rate and by constructing or financing some of the facilities, especially pipelines, used by local agencies for their conjunctive management projects. See Metropolitan Water District of Southern California (1996) and Association of Ground Water Agencies (2000).

Chapter 5. Arizona

1. However, recent and successful efforts by Native Americans to enforce their surface water rights, which are reserved under federal law, have had indirect effects on conjunctive management. Newly enforced Native American claims represent a substantial demand on surface water that must be accommodated. Many more Native American claims remain in adjudication proceedings, still unresolved. Accommodating Native American claims leaves less water available for non-Native water users, forcing such users to search for alternative sources of water, such as the water that conjunctive management would make available.

2. SRP allocations of the Salt and Verde Rivers are determined by the Kent Decree. The only SRP shareholders authorized to receive direct or "normal" flows from these rivers are those designated with "Class A" lands. These flows are subject to the prior appropriation doctrine.

3. The legislature later forbade the transport of water away from groundwater basins across the state; see Arizona Revised Statutes, Chapter 45, Article 8, §45-544.

4. Citizens receiving CAP water switched en masse to bottled water and filtration systems. A University of Arizona Water Resources Research Center study found that "customers switched from groundwater to CAP water increased their bottled water usage more than ten fold, compared to a tripling in the rate of bottled water usage for those kept on groundwater" (Gelt et al. 2001).

5. The ability of municipalities to use municipal effluent as they choose has been raised in court. In the 1989 Arizona Supreme Court case Arizona Public Service Co. v. Long, the cities of Phoenix and Tolleson were sued by two downstream appropriators when the cities began to sell their treated effluent to several utilities. In the 1970s and 1980s, the cities had been releasing their treated effluent into the Salt River and it was then used by ranchers downstream with rights to Salt River waters. The court ruled that the cities have the right to sell effluent, but if they choose to release the effluent into the

stream bed it becomes surface water that is subject to the prior appropriation doctrine. (See APS v. Long, 160 Ariz. 429, 773 P2d 988.)

6. Thirteen of the sixteen in-lieu projects were located in irrigation districts or on individual farms.

Chapter 6. Colorado

1. Historically, Colorado's Supreme Court refused to sanction a nonconsumptive instream use. It had, however, let stand a narrowly worded statute allowing the State, through the Colorado Water Conservation Board, to appropriate nonappropriated water to maintain minimum stream flows [Colorado River Water Conservation Board v. Colorado Water Conservation Board, 197 Colo. 469, 594 P.2d 570 (1979)]. In 1992, however, the Supreme Court approved an in-channel recreational water right claimed by Fort Collins [City of Thorton v. City of Fort Collins, 830 P.2d 915 (Colo. 1992)]. After the decision, other municipalities began to file applications in water court for recreational water rights to be used to support kayak courses. In 2001, the legislature adopted a law that recognized recreational in-channel water right, but limited it to counties, municipalities, water districts, water conservancy districts, and water conservation districts (Colo. Rev. Stat. §37-92-103). The law established an administrative procedure whereby the Colorado Water Conservation Board must first review such applications to consider whether the application meets the requirements of the law before the application is heard in water court (Colo. Rev. Stat. §37-92-305).

2. Originally, water courts were county district courts, which would hear claims for appropriations situated within the county. Any appropriator potentially affected by the claim, and whose rights had not been adjudicated, would be joined to the proceedings. Adjudicated rights holders would not be joined because their rights were senior and protected under existing laws (Vranesh 1987, 380).

3. Applicants bear the burden of demonstrating that their applications will not injure any existing water rights. Referees may require applicants to modify their claims to avoid injury to existing appropriations.

4. Initially, the state engineer required GASP to provide an amount of water equal to 5 percent of the amount of water pumped by its members' wells. Also, GASP was allowed to drill wells relatively close to the South Platte River, near the canals of the most senior appropriators. Instead of the senior appropriators making a call on the river, GASP would turn on its wells and divert water into the seniors' canals to satisfy their water demands. The wells would not affect the river flow until winter, when the South Platte River carries enough water for all rights to be satisfied. Over time, in response to protests from a variety of senior water-rights holders, the state engineer has ceased allowing GASP to pump groundwater and place it in canals to quiet calls on the river. Instead, the state engineer required GASP to provide additional water supplies, roughly, an amount equal to 30 percent of the water pumped by its members' wells.

5. Prior to the adoption of the Arkansas River Compact in 1948, approximately 700 wells pumped 15,000 acre-feet of water annually in Colorado. By 1995, when the case was decided, approximately 2,800 wells pumped about 150,000 acre-feet of water annually (District Court, Water Division 2, Case No. 95CW211, 1996).

6. Return flow water is the water not consumed by plants, which seeps into the ground and returns to the stream or river.

7. These generalizations apply to tributary groundwater. As noted earlier in the chapter, nontributary groundwater and groundwater in designated basins are managed as if they were not connected with surface supplies and do not affect surface water users.

Chapter 7. Tracing and Comparing Institutional Effects

1. In fact, in the 1970s, Arizona faced a problem precisely the opposite of that faced by Colorado. While Colorado struggled to incorporate groundwater within the prior appropriation doctrine, Arizona faced its first large-scale surface water adjudication—the Gila River adjudication. It struggled with incorporating native surface water governed by the prior appropriation doctrine within a water setting dominated by groundwater and project water, neither of which is governed by prior appropriation.
2. This is not the case, however, outside the AMAs. Most of Arizona's large municipal and agricultural areas lie within the AMAs, although some communities outside of the AMAs are beginning to face groundwater overdraft problems, and active discussion of whether and how to extend the state's water management framework beyond the AMAs is under way.
3. California water providers have experimented more extensively than their Arizona counterparts with alternative recharge techniques such as injection wells to prevent seawater intrusion and controlled releases of dams into streambeds. This difference, especially the use of injection wells as an alternative to large percolation basins, appears to relate to the availability and cost of land in developed areas of California.

Chapter 8. Future Directions of the Diverging Streams

1. For instance, in September and October of 2003, the prospect that up to 1 million acre-feet of additional water might be pumped from and through the Delta to meet consumptive demands in central and southern California (Taugher 2003) threatened to set off a new round of conflict between north-state and south-state interests, and the discovery of mercury contamination in the soil of thousands of acres that had been purchased by CALFED participants for a Delta wetlands restoration project (Leavenworth 2003) held the potential to postpone if not derail that element of the complex of solutions being planned for the Bay–Delta.
2. Napa Citizens for Honest Government v. Napa County Board of Supervisors, 110 Cal.Rptr.2d 579 (Ct. App. 2001).
3. The Groundwater Replenishment System (GWRS) is a joint project of the Orange County Water District and the Orange County Sanitation District, which will intercept 100,000 acre-feet per year of wastewater that would otherwise have been treated and discharged to the ocean, purify it through a combination of microfiltration, reverse osmosis, and ultraviolet disinfection, and then recharge it to the Orange County groundwater basin. There the treated water will have an expected residency time of two to five years before it would reach wells and be extracted for reuse, and only then after blending with other water supplies and further pre-tap treatment. GWRS is, therefore, both a water reuse and a conjunctive management project. In October 2003, the California

Department of Health Services issued a finding that the water produced by the GWRS would meet state standards for use in a water-recharge project.

4. Multijurisdictional arrangements are sometimes an option for a financing mechanism. However, the issuance of bonds to those entities can be complicated when different jurisdictions have different financial capacities and debt limits. Bond agencies may perceive such arrangements to be risky, or the costs to local jurisdictions of devising and maintaining viable agreements may be too costly. Although special districts may be better suited to financing projects, Title 48 of the Arizona Revised Statutes does not provide for a special district for water infrastructure development that can encompass existing municipal, special district, and county jurisdictions.

5. Through 1997, 1,516,388 acre-feet of long-term storage credits for groundwater were held by 36 entities in the Phoenix, Tucson, and Pinal AMAs.

6. In addition to dealing with the recovery location of stored water within AMAs, Arizona will have to address the need for conjunctive management outside the AMAs. For example, the City of Sierra Vista, near the Fort Huachuca Army Base, confronts serious ecological impacts on the San Pedro River National Conservation Area caused by unlimited groundwater pumping. In 2001, the City of Sierra Vista received a permit from the state to develop a conjunctive management program, recharging up to 4,149 acre-feet of municipal effluent per year. This project and four small projects that recently received permits outside of the Phoenix AMA constitute the whole of non-AMA activity to this point.

7. In response to a Colorado Supreme Court decision that questioned the authority of the state engineer to issue substitute supply plans, the state engineer initiated a rule making process for Division 1. The new rule, modeled on the replacement plan rules of the Arkansas Basin, would establish more specific water replacement requirements for wells pumping out of priority. The rulemaking process is not expected to conclude for several months.

Chapter 9. Shaping the Future

1. Whether to prevent transfers to certain water uses, across basins, or away from instream flow protection are policy choices that must be addressed by the citizens and lawmakers of each state, region, or locality.

2. In particular, the salinity of stream flows currently recharged into groundwater basins in Southern California has become a point of serious concern, as the rising salt load is gradually degrading the quality of groundwater that millions of Southern California residents and businesses rely on.

3. It is especially important to put these considerations into place before other policy recommendations that would promote even more replenishment, such as the Governor's Water Management Commission's recommendation that the Central Arizona Groundwater Replenishment District (CAGRD) establish a "replenishment reserve" of long-term storage credits (see Megdal 2002).

4. Sophocleous (2000) has found that locally driven decision-making in Kansas has been particularly successful in devising sustainable groundwater management plans. Local groundwater management districts have been able to clarify the particular hydrologic demands of ground and surface water systems and then devise safe-yield policies to reflect the variability among different groundwater basins.

Appendix

1. Copies of the coding forms may be obtained by contacting the authors.

2. This research framework is formally known as the Institutional Analysis and Development (IAD) framework (see Ostrom 1990, 1999; Ostrom et al. 1994). The framework incorporates elements of human behavior recognized in the fields of political science, public administration, sociology, and economics in looking at the effects of institutions, choices, and action. Its emphasis on community context of action also integrates theoretical insights from the fields of history and anthropology.

3. One was the North Coast region, which is sparsely populated, contains few usable groundwater basins, and where groundwater tends to be found in numerous small volcanic rock basins that are not conducive to storage and retrieval of much water. The other was the North Lahontan region, which is the northern reach of Sierra Nevada mountains, also sparsely populated with groundwater resources not conducive to conjunctive management. Information on the regions can be found in California Department of Water Resources (1998a–c).

Bibliography

References

Abbott, P.O. (1985) *Description of Water Systems Operations in the Arkansas River Basin, Colorado.* Water Resources Investigations Report 85-4092. Lakewood, CO: U.S. Geological Survey.

Aiken, J. David (1999) Balancing Endangered Species Protection and Irrigation Water Rights: The Platte River Cooperative Agreement. *Great Plains Natural Resources Journal* 3: 19–158.

Anderson, Terry, Oscar Burt, and David Fractor (1983) Privatizing Groundwater Basins: A Model and Its Application. In *Water Rights.* Terry Anderson, ed. San Francisco, CA: Pacific Institute for Public Policy Research, pp. 223–248.

Arizona Department of Water Resources (1999a) *Phoenix Active Management Area Draft Third Management Plan: 2000–2010.* Phoenix, AZ: Arizona Department of Water Resources.

———(1999b) *Tucson Active Management Area Draft Third Management Plan: 2000–2010.* Phoenix, AZ: Arizona Department of Water Resources.

———(1999c) *Prescott Active Management Area Draft Third Management Plan: 2000–2010.* Phoenix, AZ: Arizona Department of Water Resources.

———(2001a) Inside AMAs. *water.az.gov/azwaterinfo/insideAMAs*

———(2001b) Arizona's Water Supplies and Water Demands. *water.az.gov/azwaterinfo/statewide/*

———(2001c) Semi–Annual Status Report of Permitted Facilities, June 30th 2001. Phoenix, AZ: Arizona Department of Water Resources.

———(2003) Assured Water Supply Program Brochure. *http://www.water.az.gov/watermanagement/Content/Forms/AWSBrochure.pdf* (Accessed October 27, 2003).

Arizona Water Banking Authority (1998) *Annual Plan of Operation.* Phoenix, AZ: AWBA (January).

———(1999) *Annual Plan of Operation.* Phoenix, AZ: Arizona Water Banking Authority (January).

———(2000) *Annual Plan of Operation.* Phoenix, AZ: Arizona Water Banking Authority (January).

————(2001) *Annual Plan of Operation.* Phoenix, AZ: Arizona Water Banking Authority (January).

Arizona Water Banking Authority Study Commission (1998) *Arizona Water Banking Authority Study Commission Final Report.* Phoenix, AZ: Arizona Department of Water Resources (December).

Association of Ground Water Agencies (1998) Conjunctive Use Issue Paper. Prepared for the AGWA/MWD/NWRI Nominal Group Technique Workshop on Conjunctive Use. Pomona, CA (May).

————(2000) *Groundwater and Surface Water in California: A Guide to Conjunctive Use.* Fountain Valley, CA: Association of Ground Water Agencies.

Atwater, Richard and William Blomquist (2002) Rates, Rights and Regional Planning in the Metropolitan Water District of Southern California. *Journal of the American Water Resources Association.* 38(5): 1195-1205 .

Bachman, Steve, Carl Hauge, Kevin Neese, and Anthony Saracino (1997) *California Groundwater Management.* Sacramento, CA: Groundwater Resources Association of California.

Banks, Harvey O. (1953) Utilization of Underground Storage Reservoirs. *Transactions, American Society of Civil Engineers* 118: 220–234.

Bates, Robert H. (1989) *Beyond the Miracle of the Market.* New York: Cambridge University Press.

Bish, Robert (1977) Environmental Resource Management: Public or Private? In *Managing the Commons.* Garrett Hardin and John Baden, eds. San Francisco, CA: W.H. Freeman, pp. 217–228.

Blomquist, William (1992) *Dividing the Waters: Governing Groundwater in Southern California.* San Francisco, CA: ICS Press.

————(1998) *Water Security and Future Development in the San Juan Basin: The Role of the San Juan Basin Authority.* Fountain Valley, CA: National Water Research Institute.

Bruggink, Thomas H. (1992) Third Party Effects of Groundwater Law in the United States. *American Journal of Economics and Sociology* 51(2): 202–222.

Byers, Jack (2002) Drought Hits Colorado. *StreamLines: Quarterly Newsletter of the Office of the State Engineer* 16(2): 1–2.

California Department of Water Resources (1993) *State Drought Water Bank Program Environmental Impact Report.* Sacramento, CA; California Department of Water Resources (November).

————(1994) *California Water Plan Update.* Bulletin 160-93. Executive Summary. Sacramento, CA: California Department of Water Resources.

————(1995) *Management of the State Water Project.* Bulletin 132-94. Sacramento, CA: California Department of Water Resources.

————(1998a) *California Water Plan Update.* Bulletin 160-98. Executive Summary. Sacramento, CA: California Department of Water Resources.

————(1998b) *California Water Plan Update.* Bulletin 160-98. Volume 1. Sacramento, CA: California Department of Water Resources.

California Water Resources Control Board (1994) The Water Right Process. Sacramento, CA: California Water Resources Control Board, 4 pp.

Central Arizona Water Conservation District (1999) *Annual Report.*

———— (2001) Timeline. *www.cap–az.com/about/history/timeline*

———— (2003a) Rates. *www.cap–az.com/management/rates*

———— (2003b) CAGRD, Executive Summary. *www.cap–az.com/* (Accessed October 28, 2003).

Checchio, E. (1988) *Water Farming: The Promise and Problems of Water Transfers in Arizona.* Water Resources Research Center Issue Paper No. 4. Tucson, AZ: University of Arizona Water Resources Research Center.

CIC Research, Inc. (1999) The Economic Impact on San Diego County of Three Levels of Water Delivery: 80, 60 or 40 Percent Occurring for Two or Six Months. Phase II Executive Summary. Report to the San Diego County Water Authority (October), 14 pp.

Colby, Bonnie G. (1989) Estimating the Value of Water in Alternative Uses. *Natural Resources Journal* 29(2): 511–527.

Colorado Department of Agriculture (2000) Colorado Agricultural Statistics, 2000 preliminary and 1999 revised. Lakewood, CO: Colorado Agriculture Statistics Service. *http://www.nass.usda.gov/co/*

Colorado Division of Water Resources (2000a) Cumulative Yearly Statistics, Water Year 2000. Denver, CO: Colorado Division of Water Resources.

———(2000b) Colorado Historic Average Annual Stream Flows. Denver, CO: Colorado Division of Water Resources.

———(2000c) Transmountain Diversions: Office of the State Engineer. Denver, CO: Colorado Division of Water Resources.

Colorado Groundwater Commission (2001) Designated Basins and Groundwater Management Districts. *http://www.water.state.co.us/cgwc*

Colorado State Engineer's Office (1996) Amended Rules and Regulations Governing the Diversion and Use of Tributary Groundwater in the Arkansas River Basin. Denver, CO: Colorado State Engineer's Office.

Colorado Stream Lines (2001a) Lower South Platte River Water User Activities. *StreamLines: Quarterly Newsletter of the Office of the State Engineer.* 15(2): 2.

———(2001b) Arkansas River Water Banking Pilot Project. *StreamLines: Quarterly Newsletter of the Office of the State Engineer.* 15(2): 3.

Colorado Water Protective and Development Association (1998) Augmentation Plan. (mimeograph).

Commons, John R. (1968) *Legal Foundations of Capitalism.* Madison, WI: University of Wisconsin Press.

Conkling, Harold (1946) Utilization of Ground-Water Storage in Stream System Development. *Transactions, American Society of Civil Engineers* 111: 275–354.

Cooperative Agreement for Platte River Research and Other Efforts Relating to Endangered Species Habitats Along the Central Platte River, Nebraska (July 1997).

Cuzan, Alfred G. (1979) A Critique of Collectivist Water Resources Planning. *Western Political Quarterly* 32(3): 320–326.

Denver Water Board (2003) Who We Are. *www.denverwater.org/whoweare/* (Accessed October 20, 2003).

DeYoung, Tim and Hank Jenkins-Smith (1989) Privatizing Water Management: The Hollow Promises of Private Markets. In *Water and the Future of the Southwest.* Zachary A. Smith, ed. Albuquerque, NM: University of New Mexico Press, pp. 213–231.

Fischer, Ward H. and Steven B. Ray (1978) *A Guide to Colorado Water Law.* Fort Collins, CO: Colorado State University/Colorado Water Resources Research Institute.

Fisher, Anthony, David Fullerton, Nile Hatch, and Peter Reinelt (1995) Alternatives for Managing Drought: A Comparative Cost Analysis. *Journal of Environmental Economics and Management* 29(3): 304–320.

Fradkin, Philip (2003) Water Pact May Soon Evaporate. *The Los Angeles Times* (October 21).

Gardner, Roy, Michael R. Moore, and James M. Walker (1997) Governing a Groundwater Commons: A Strategic and Laboratory Analysis of Western Water Law. *Economic Inquiry* 35(2): 218–234.

Gelt, Joe, et al. (2001) *Water in the Tucson Area: Seeking Sustainability.* Water Resources Research Center, College of Agriculture, The University of Arizona, Tucson, AZ.

Glen Canyon Institute (2001) Facts About Reservoir Powell. *www.glencanyon.org/background/factsdam.htm*

Glennon, Robert J. and Thomas Maddock III (1994) In Search of Subflow: Arizona's Futile Effort to Separate Groundwater from Surface Water. *Arizona Law Review* 36(3): 567–610.

Governor's Water Management Commission (2001) *Governor's Water Management Commission Final Report.* Phoenix, AZ: State of Arizona.

Gregg, Frank, Stephen M. Born, William B. Lord, and Marvin Waterstone (1991) *Institutional Response to a Changing Water Policy Environment.* Final Report, U.S. Geological Survey Grant No. 14-08-0001-G1639. Tucson, AZ: Arizona Water Resources Research Center.

Hardin, Garrett (1968) The Tragedy of the Commons. *Science* 162: 1243–1248.

Hauge, Carl J. (1992) The Importance of Ground Water in California. In *Changing Practices in Ground Water Management—The Pros and Cons of Regulation.* Proceedings of the 18th Biennial Conference on Ground Water. Report No. 77. Riverside, CA: University of California Water Resources Center, pp. 15–30.

———— (1994) The Physical Dimensions of Sustainability. *Proceedings, Nineteenth Biennial Conference on Ground Water.* Water Resources Center Report No. 84. Davis, CA: University of California Centers for Water and Wildland Resources, pp. 9–13.

Heikkila, Tanya (2001) Managing Common-Pool Resources in a Public Service Industry: The Case of Conjunctive Water Management. Ph.D. dissertation, School of Public Administration and Policy, University of Arizona.

Huffman, James L. (1986) Allocating Water to Instream Uses: Private Alternatives. In *Water Resources Law: Proceedings of the National Symposium on Water Resources Law.* St. Joseph, MI: American Society of Agricultural Engineers.

Hundley, Norris, Jr. (2001) *The Great Thirst: Californians and Water, A History.* Berkeley, CA: University of California Press.

Ingram, Helen M. and Cy Oggins (1992) The Public Trust Doctrine and Community Values in Water. *Natural Resources Journal* 32(3): 515–537.

Ingram, Helen M. et al. (1984) Guidelines for Improved Institutional Analysis in Water Resources Planning. *Water Resources Research* 20(3): 323–334.

Johnson, James W. and Margaret Gallogly (2001) Arizona and Nevada Sign Interstate Water Banking Agreement. *Water Law Newsletter* 34(3): 1+

Johnson, Kyle C. (1992) Letting the Free Market Distribute Environmental Resources. *William and Mary Journal of Environmental Law* 17(1): 79–102.

Kepler, Keith (2003) Second Year Drought Affects Arkansas River Basin Well Users. *StreamLines: Quarterly Newsletter of the Office of the State Engineer* 17(1): 2–4.

Kiser, Larry and Elinor Ostrom (1982) The Three Worlds of Action: A Metatheoretical Synthesis of Institutional Approaches. In *Strategies of Political Inquiry.* Elinor Ostrom, ed. Beverly Hills: Sage, pp. 179–222

Kondolf, G. Mathias (1994) Surface–Ground Water Interactions: Some Implications for Sustainability of Ground Water Resources. *Proceedings, Nineteenth Biennial Conference on Ground Water.* Water Resources Center Report No. 84. Davis, CA: University of California Centers for Water and Wildland Resources, pp. 133–142.

Knapp, Keith C. and Lars J. Olson (1995) The Economics of Conjunctive Groundwater Management with Stochastic Surface Supplies. *Journal of Environmental Economics and Management* 28(3): (May) 340–356.

Krane, Dale, Platon N. Rigos, and Melvin B. Hill, Jr. (2001) *Home Rule in America: A Fifty-State Handbook.* Washington, DC: CQ Press.

Krautkraemer, John (1992) Pros and Cons of Ground Water Regulation. In *Changing Practices in Ground Water Management—The Pros and Cons of Regulation.* Proceedings of the

18th Biennial Conference on Ground Water. Report No. 77. Riverside, CA: University of California Water Resources Center, pp. 151–153.

Leavenworth, Stuart (2003) Toxic Dilemma. *The Sacramento Bee.* (October 20).

Leshy, John D. and James Belanger (1988) Arizona Law: Where Ground and Surface Water Meet. *Arizona State Law Journal.* 20(3): 657–748.

Littleworth, Arthur L. and Eric L. Garner (1995) *California Water.* Point Arena, CA: Solano Press Books.

Livingston, Marie L. (1993) Designing Water Institutions: Market Failure and Institutional Response. Policy Research Working Paper No. 1227. Washington, DC: The World Bank.

Long's Peak Working Group on National Water Policy (1994) America's Waters: A New Era of Sustainability. *Environmental Law* 24(1): 125–144.

Lord, William B. (1984) Institutions and Technology: Keys to Better Water Management. *Water Resources Bulletin* 20(5): 651–656.

Lower Colorado River Multispecies Conservation Plan (2001) The Lower Colorado River Multi-species Conservation Program. *www.lcrmscp.org/Description.html*

MacDonnell, Lawrence (1988) Colorado's Law of "Underground Water:" A Look at the South Platte Basin and Beyond. *University of Colorado Law Review* 59(3): 579–625.

Mann, Dean (1963) *The Politics of Water in Arizona.* Tucson, AZ: University of Arizona Press.

Matthews, Christine A. (1991) Using Ground Water Basins as Storage Facilities in Southern California. *Water Resources Bulletin* 27(5): 841–847.

Megdal, Sharon (2002) The Central Arizona Groundwater Replenishment District—The Need for Some Fine Tuning. *Arizona Water Resource* 10(5): 11.

Metropolitan Water District of Southern California (1996) *Southern California's Integrated Water Resources Plan.* Report No. 1107. Executive Summary. Los Angeles, CA: Metropolitan Water District of Southern California.

Metzger, Philip C. (1988) Protecting Social Values in Western Water Transfers. *American Water Works Association Journal* 80(3): 58–65.

Mills, William R. Jr. (1994) Recharge with Recycled Water: A Global Perspective. *Proceedings, Nineteenth Biennial Conference on Ground Water.* Water Resources Center Report No. 84. Davis, CA: University of California Centers for Water and Wildland Resources, pp. 31–34.

Morehouse, Barbara J. (2002) Integrating Climate into Water Policy. *Southwest Hydrology* 1(2): 16+

Morrison, Jason I., Sandra L. Postel, and Peter H. Gleick (1996) *The Sustainable Use of Water in the Lower Colorado River Basin.* Oakland, CA: Pacific Institute for Studies in Development, Environment, and Security.

National Research Council (1997) *Valuing Ground Water: Economic Concepts and Approaches.* Washington, DC: National Academy Press.

National Research Council, Water Science and Technology Board (2001) *Envisioning the Agenda for Water Resources Research in the Twenty-First Century.* Washington, DC: National Academy Press.

National Water Research Institute (1998) *Conjunctive Use Water Management Program.* Workshop Report. Fountain Valley, CA: National Water Research Institute.

Natural Heritage Institute (1997) *Feasibility Study of a Maximal Groundwater Banking Program for California.* Working Draft. (May 8), 67 pp.

North, Douglass C. (1990) *Institutions, Institutional Change, and Economic Performance.* New York: Cambridge University Press.

Northern Colorado Water Conservancy District (2001) Colorado–Big Thompson Project. *http://www.ncwcd.org/project&features/cbt_main.htm*

Oakerson, Ronald (1999) *Governing Local Public Economies.* San Francisco, CA: ICS Press.

O'Connell, Maureen (2001a) Thousands Still Seek Payments. *Arizona Daily Star* (April 29), p. A6.

————— (2001b) CAP: Round Two. *Arizona Daily Star* (April 29), p. A1.

Oshio, Kazuto (1997) Who Pays and Who Benefits? Metropolitan Water Politics in Twentieth-Century Southern California. *The Japanese Journal of American Studies* 8: 63–89.

Ostrom, Elinor (1998) Reflections on the Commons. In *Managing the Commons,* 2nd edit. John A. Baden and Douglas S. Noonan, eds. Bloomington and Indianapolis, IN: Indiana University Press, pp. 95–116.

—————(1990) *Governing the Commons.* Cambridge: Cambridge University Press.

Paddock, William (2003) 2003 Colorado Water Legislation. *Water Law Newsletter* 36(2): 8–9.

Paddock, William and Mary Mead Hammond (2002) Colorado: Instream Flow Rights. *Water Law Newsletter* 35(2): 7.

Perry, Tony (2003) Imperial Water Deal Completed. *The Los Angeles Times* (October 3).

Pianen, Eric (2002) Army Corps Tries to Solve Dispute Over Water Levels. *The Washington Post* (May 26).

Porter, Mary Jean (2001) Southern Colorado Water Bank Plan Draws Questions, Concerns. *The Pueblo Chieftain* (October 8), p. 1.

Postel, Sandra L., Jason I. Morrison, and Peter H. Gleick (1998) Allocating Fresh Water to Aquatic Ecosystems: The Case of the Colorado River Delta. *Water International* 23(3): 119–125.

Provencher, Bill (1993) A Private Property Rights Regime to Replenish a Groundwater Aquifer. *Land Economics* 69(4): 325–340.

Radosevich, G.E., K.C. Nobe, D. Allardice, and C. Kirkwood (1976) *Evolution and Administration of Colorado Water Law: 1876–1976.* Fort Collins, CO: Water Resources.

Reisner, Marc (1994) Deconstruction in the Arid West: Close of the Age of Dams. *West–Northwest* 1(1): 1–11.

Reisner, Marc and Sarah Bates, eds. (1990) *Overtapped Oasis: Reform or Revolution for Western Water.* Washington, DC: Island Press.

Robie, Ronald B. (2001) New Legislation Requires Consideration of Water Supplies for New Developments. *Water Law Newsletter* 34(3): 6–7.

Rose, Carol (1994) *Property and Persuasion: Essays on the History, Theory, and Rhetoric of Ownership.* Boulder, CO: Westview Press.

Rouse, Karen (2001) Cost of Water Rises as Aquifers Shrink. *The Denver Post* (June 24), p. A–01.

Runge, C.F. (1984) Institutions and the Freerider: The Assurance Problem in Collective Action. *Journal of Politics* 46(1): 154–181.

Salt River Project (1999) *1997–1998 Salt River Project Annual Report.* Tempe, AZ: Salt River Project.

—————(2001) Water. *www.srpnet.com/water/*

Sax, Joseph L. (1994) Understanding Transfers: Community Rights and the Privatization of Water. *West–Northwest* 1(1): 13–16.

Schlager, Edella (1999) Colorado Water Law as Customary Law: The South Platte and Arkansas River Basins. Paper prepared for the Workshop on the Workshop II, Bloomington, Indiana, June 9–12, 1999.

Schlager, Edella and Elinor Ostrom (1992) Property Rights Regimes and Natural Resources: An Empirical Analysis *Land Economics* 68(3): 249–262.

Smith, Jerd (2002) A War About Water. *Rocky Mountain News.* (November 18), p. 4A.

—————(2003) Ruling, Law Will Allow Farms to Pump Water. *Rocky Mountain News* (May 1), p.10A.

Smith, Zachary B., ed. (1989) *Water and the Future of the Southwest.* Albuquerque, NM: University of New Mexico Press.

Solley, Wayne B, Robert R. Pierce, and Howard A. Perlman (1993) *Estimated Water Use in the United States 1990.* United States Geological Survey Circular 1081. Washington, DC: U.S. Government Printing Office.

———(1998) *Estimated Water Use in the United States 1995.* United States Geological Survey Circular 1200. Washington, DC: U.S. Government Printing Office.

Sophocleous, Mario (2000) From Safe Yield to Sustainable Development of Water Resources—the Kansas Experience. *Journal of Hydrology* 235: 27–43.

Supalla, Raymond, Bettina Klaus, Osei Yeboah, and Randall Bruins (2002) A Game Theory Approach to Deciding Who Will Supply Instream Flow Water. *Journal of the American Water Resources Association* 38(4): 959–966.

Tang, Shui Yan (1992) *Institutions and Collective Action: Self-Governance in Irrigation.* San Francisco, CA: ICS Press.

Taugher, Mike (2002) Reservoir Expansion Study Unsettled. *Contra Costa Times.* (August 14).

———(2003) Most Wanted: Delta Water. *Contra Costa Times* (September 30).

Thomas, R.O. (1955) General Aspects of Planned Ground-Water Utilization. *Proceedings, American Society of Civil Engineers* 81(Item 706). 11 pp.

Tobin, Mitch (2001a) Answers to Common Questions on CAP Water. *Arizona Daily Star* (May 3), p. A1.

——— (2001b) Sales of Bottled Water Slower This Time Around. *Arizona Daily Star* (May 2), p. A1.

Todd, David Keith and Iris Priestaf (1997) Role of Conjunctive Use in Groundwater Management. *Proceedings of the American Water Resources Association Conference, Symposium on Conjunctive Use of Water Resources: Aquifer Storage and Recovery.* (October), pp. 139–145.

Tucson AMA Safe-Yield Task Force (2000) Financing Water Infrastructure. Draft Issue Outline, July 3. Tucson, AZ: Arizona Department of Water Resources, Tucson Active Management Area.

U.S. Advisory Commission on Intergovernmental Relations (1991) *Coordinating Water Resources in the Federal System: The Groundwater–Surface Water Connection.* Report No. A–118. Washington, DC: U.S. Advisory Commission on Intergovernmental Relations.

U.S. Bureau of Reclamation (2000) Blanket Approval of Temporary Transfers and Exchanges of Project Water Between South of Delta CVP Contractors. Draft Environmental Assessment. (January) Fresno, CA: U.S. Bureau of Reclamation Mid-Pacific Region. 16 pp.

U.S. Census Bureau (2001) *State and County Quick Facts. http://quickfacts.census.gov*

U.S. Department of Agriculture, Economic Research Service (1998) *Farm Business Economic Report, 1997* (ECI–1998).

———(2001) 1997 Census of Agriculture Highlights. *www.nass.usda.gov/census/census97/highlights/ag–state.htm*

Vranesh, George (1987) *Colorado Water Law.* 3 vols. Boulder, CO: Natural Resources Law Center.

Water Resources Research Center (1999) *Water in the Tucson Area: Seeking Sustainability.* Tucson, AZ: Water Resources Research Center, University of Arizona.

Water Tech Online (2003) California Dam Decision Signals Major Water Policy Shift. *Southwest Hydrology* 2(5): 10.

Weissenstein, Michel (2000) Six Groups Protest Water Release Plan. *Las Vegas Review–Journal* (October 31).

Western Regional Climate Center (2001) *Precipitation Maps of the Western U.S.* *http://www.wrcc.dri.edu/pcpn/*

Western States Water Council (Anthony G. Willardson, Principal Investigator) (1990) *Groundwater Recharge Projects in the Western United States: Economic Efficiency, Financial Feasibility, and Legal/Institutional Issues.* Report to the Bureau of Reclamation and the Department of the Interior. Midvale, UT: WSWC.

Wilkinson, Charles F. (1988) To Settle a New Land: An Historical Essay on Water Law and Policy in the American West and in Colorado. In *Water and the American West: Essays in Honor of Raphael J. Moses.* David H. Getches, ed. Boulder, CO: Natural Resources Law Center, pp. 1–17.

Woodhouse, Betsy (2002) Climatic Effects on Groundwater Conditions in the Southwest. *Southwest Hydrology* 1(2): 26+

Young, Oran R. (1982) *Resource Regimes: Natural Resources and Social Institutions.* Berkeley, CA: University of California Press.

Cases Cited

State Cases: *Arizona*

APS v. Long, 160 Ariz. 429, 773 P2d 988.

State Cases: *California*

Baldwin v. County of Tehama (1994).

County of Inyo v. Los Angeles Department of Water and Power, 32 Cal. App. 3d 795 (1973).

Napa Citizens for Honest Government v. Napa County Board of Supervisors, 110 Cal.Rptr.2d 579 (Ct. App. 2001).

National Audubon Society v. Superior Court, 33 Cal.3d 419 (1983).

State Cases: *Colorado*

City of Thornton v. City of Fort Collins, 830 P.2d 915 (Colo. 1992).

Colorado River Water Conservation Board v. Colorado Water Conservation Board, 197 Colo. 469, 594 P.2d 570 (1979).

Concerning the Amended Rules and Regulations Governing the Diversion and Use of Tributary Groundwater in the Arkansas River Basin, Colorado, Case No. 95CW211, District Court, Water Division 2.

In re Arkansas River, 195 Colo. 557, 581 P.2d 293, 1978.

In re the Proposed Amended Rules and Regulations Governing the Diversion and Use of Tributary Ground Water in the S. Platte River Basin, No. 02-CW-108 D.C. Water Div. No.1. Colo. Dec. 23, 2002.

Empire Lodge Homeowners' Ass'n v. Moyer, 39 P.3d 1139 (Colo. 2001).

Fellhauer v. People, 167 Colo. 320, 447 P.2d 986 (1969).

Simpson v. Bijou Irrigation Co., 69 P.3d 50(Colo.2003).

Federal Cases

Kansas v. Colorado, 514 U.S. 673 (1995).

Other Source Materials

Abbott, Carl, Stephen J. Leonard, and David McComb (1994) *Colorado: A History of the Centennial State,* 3rd edit. Niwot, CO: University Press of Colorado.

Anderson, James W. (1989) Some Thoughts on Conjunctive Use of Ground Water in California. *Western State University Law Review* 16(2): 559–589.

Andrews, Barbara T. and Sally K. Fairfax (1984) Groundwater and Intergovernmental Relations in the Southern San Joaquin Valley of California: What Are All These Cooks Doing to the Broth? *University of Colorado Law Review* 55(2): 145–271.

Arizona Department of Water Resources (1991) *Second Management Plan: Tucson Active Management Area.* Phoenix, AZ: Arizona Department of Water Resources.

——— (1996) *Regional Recharge Committee: Technical Report.* Tucson, AZ: ADWR Tucson Active Management Area.

———(1997) *Arizona Water Information: Statewide Surface Water Management.* http://www.adwr.state.az.us/azwaterinfo/index.html

——— (1998) Sahuarita–Green Valley Area: Central Arizona Project Water Use Feasibility Analysis and Delivery System Optimization Study. *Task 4 Technical Memorandum: Financial Feasibility Analysis.* Tucson, AZ: Arizona Department of Water Resources, Tucson Active Management Area.

Arizona Senate Staff (2001) Fact Sheet for S.C.M. 1002. *http://www.azleg.state.az.us/legtext/45leg/1r/bills/scm1002o.htm*

Arizona Water Banking Authority (1998) *Facility Plan: Tucson Active Management Area.* Phoenix, AZ: Arizona Water Banking Authority (September).

Bates, Sarah, David Getches, Lawrence MacDonnell, and Charles Wilkinson (1993) *Searching Out the Headwaters: Change and Rediscovery in Western Water Policy.* Washington, DC: Island Press.

Birdlebough, Stephen C. and Alfred Wilkins (1971) Legal Aspects of Conjunctive Use in California. In *California Water: A Study in Resource Management.* David Seckler, ed. Berkeley, CA: University of California Press, pp. 263–270.

Bittinger, Morton (1964) The Problem of Integrating Ground Water and Surface Water Use. *Ground Water* 2(3): 33–38.

Blomquist, William, Tanya Heikkila, and Edella Schlager (2001) Institutions and Conjunctive Water Management in Three Western States. *Natural Resources Journal.* 41(3):653-683.

Bookman–Edmonston Engineering, Inc. (1999) Drought Benefits of Southern California Ground Water Basins Within the Metropolitan Water District Service Area. Discussion paper prepared for the Association of Ground Water Agencies. Glendale, CA (November).

Born, Stephen, M. ed. (1989) *Redefining National Water Policy: New Roles and Directions.* Bethesda, MD: American Water Resources Association.

Bowles, Jennifer (2003) Fine Sought Against MWD. *Riverside Press–Enterprise.* (September 20).

Briggs, Philip C. (1983) Ground-Water Management in Arizona. *Journal of Water Resources Planning and Management* 109(3): 195–202.

Brown, Linton A. (1994) Conjunctive Use: Problems and Advantages. *Proceedings, Nineteenth Biennial Conference on Ground Water.* Water Resources Center Report No. 84. Davis, CA: University of California Centers for Water and Wildland Resources, pp. 25–30.

Burges, Stephen J. and Reza Marnoon (1975) *A Systematic Examination of Issues in Conjunctive Use of Ground and Surface Waters.* Water Resources Information System Technical Bulletin No. 7. Olympia, WA: State of Washington Department of Ecology.

California Department of Water Resources (1994a) *California Water Plan Update.* Bulletin 160-93. Volume 1. Sacramento, CA: California Department of Water Resources.

——— (1994b) *California Water Plan Update.* Bulletin 160-93. Volume 2. Sacramento, CA: California Department of Water Resources.

——— (1998) *California Water Plan Update.* Bulletin 160-98. Volume 2. Sacramento, CA: California Department of Water Resources.

——— (2003) *Draft California Water Plan Update.* www.water.ca.gov/b160/ Stakeholder_Briefing_Draft (Accessed October 7, 2003).

Case, Pamela J. and Gregory Alward (1997) Patterns of Demographic, Economic and Value Change in the Western United States: Implications for Water Use and Management. In *Report to the Western Water Policy Review Advisory Commission.* Springfield, VA: National Technical Information Service.

Challen, Ray (1995) Regulation, Imperfect Markets and Transaction Costs: The Elusive Quest for Efficiency in Water Allocation. In *Handbook of Environmental Economics,* D. Bromley, ed. Oxford: Blackwell.

———(2000) *Institutions, Transaction Costs and Environmental Policy: Institutional Reform for Water Resources.* Cheltenham, UK: Edward Elgar.

CIC Research, Inc. (1999) The Economic Impact on San Diego County of Three Levels of Water Delivery: 80, 60, or 40 Percent Occurring for Two Months of Six Months. Phase II Executive Summary. Prepared for San Diego County Water Authority. (October 19), 14 pp.

Colorado Water Resources Research Institute (1991) *Colorado Citizens Water Handbook— Colorado Water: The Next 100 Years.* Information Series No. 67. Boulder, CO: Colorado State University.

Connal, Desmond D., Jr. (1982) A History of the Arizona Groundwater Management Act. *Arizona State Law Journal* 1982(2): 313–343.

Ditmer, Joanne (1997) Transferring Water Rights All Wet? *The Denver Post* (October 19), p. F-2.

Dycus, J. Stephen (1984) Development of a National Groundwater Protection Strategy. *Boston College Envirnmental Affairs Law Review* 11(2):211–271.

Dzurik, Andrew A. (1990) *Water Resources Planning.* Savage, MD: Rowman & Littlefield.

Eden, Susanna (1990) *Integrated Water Management in Arizona.* Issue Paper Number Five. Tucson, AZ: Arizona Water Resources Research Center.

El-Ashry, Mohamed T. and Diana C. Gibbons (1986) *Troubled Waters: New Policies for Managing Water in the American West.* Washington, DC: World Resources Institute.

Environmental and Energy Study Institute (1993) *New Policy Directions to Sustain the Nation's Water Resources.* Washington, DC: Environmental and Energy Study Institute.

Farrow, Ross (2002) District Prepares for Recharge Project. *Lodi News–Sentinel.*

———(2003) Kern County Knows Groundwater Banking. *Lodi News–Sentinel.* (July 14).

Florkowski, Joe (2003a) New Wells to Improve Water Quality. *Inland Valley Daily Bulletin* (September 11).

———(2003b) Early Talks Held for Water Storage Plan. *Inland Valley Daily Bulletin.* (October 6).

Fradkin, Philip (1981) *A River No More: The Colorado River and the West*. New York: Knopf.

Fram, Miranda S., Brian A. Bergamaschi, Kelly D. Goodwin, Roger Fujii, and Jordan F. Clark (2003) *Processes Affecting the Trihalomethane Concentrations Associated with the Third Injection, Storage, and Recovery Test at Lancaster, Antelope Valley, California*. Water Resources Investigations Report No. 034062. United States Geological Survey. *water.usgs.gov/pubs/wri/wri034062* (Accessed September 18, 2003).

Garner, Eric L. and Janice L. Weis (1991) Coping with Shortages: Managing Water in the 1990s and Beyond. *Natural Resources and Environment* 5(4): 26+

Garner, Eric L., Michelle Ouelette, and Richard L. Sharff, Jr. (1994) Institutional Reforms in California Groundwater. *Pacific Law Journal* 25(3): 1021–1052.

Getches, David H. (1985) Controlling Groundwater Use and Quality: A Fragmented System. *Natural Resources Lawyer* 17(4):623–645.

Glaser, Harold T., Donald E. Evenson, and Mark J. Wildermuth (1997) Conjunctive Use of Groundwater and Imported Surface Waters in Southern California. In *Conjunctive Use of Water Resources: Aquifer Storage and Recovery*. American Water Resources Association (October), pp. 111–120.

Gleason, Victor E. (1976) Water Projects Go Underground. *Ecology Law Quarterly* 5(4): 625–668.

Gottlieb, Robert and Margaret FitzSimmons (1991) *Thirst for Growth: Water Agencies as Hidden Government in California*. Tucson, AZ: University of Arizona Press.

Grant, Douglas L. (1987) The Complexities of Managing Hydrologically Connected Surface Water and Groundwater Under the Appropriation Doctrine. *Land and Water Law Review* 22(1): 63–95.

Groundwater Appropriators of the South Platte. Groundwater Appropriators of the South Platte River Basin, Inc. Brochure.

Harding, Sidney T. (1960) *Water in California*. Palo Alto, CA: N–P Publications.

Harrison, David L. and Gustave Sandstrom, Jr. (1971) The Groundwater–Surface Water Conflict and Recent Colorado Water Legislation. *University of Colorado Law Review* 43(1): 1–48.

Higbee, Don, Director of Lower Arkansas Water Management Association, interview in Lamar, Colorado, October 1998.

Howitt, Richard, Nancy Moore, and Rodney T. Smith (1992) *A Retrospective on California's 1991 Emergency Drought Water Bank*. Report prepared for the California Department of Water Resources. 76 pp.

Hundley, Norris, Jr. (1975) *Water and the West*. Berkeley, CA: University of California Press.

Jaquette, David L. (1978) *Efficient Water Use in California: Conjunctive Management of Ground and Surface Reservoirs*. Report R-2389-CSA/RF. Santa Monica, CA: RAND Corporation.

Johnson, Thelma A. and Helen J. Peters (1967) Regional Integration of Surface and Ground Water Resources. Presented at the Symposium of the International Association of Scientific Hydrology, Haifa, Israel.

Kahn, Jeffrey J. and Robert A. Longenbaugh (1986) The Colorado Experience in Resolving Surface–Ground Water Conflicts. In *Water Resources Law: Proceedings of the National Symposium on Water Resources Law*. St. Joseph, MI: American Society of Agricultural Engineers, pp. 76–83.

Kahrl, William L. (1982) *Water and Power: The Conflict Over Los Angeles' Water Supply in the Owens Valley*. Berkeley, CA: University of California Press.

Kennedy, David N. (1994) California's Ground Water: A Vital Resource. *Proceedings, Nineteenth Biennial Conference on Ground Water*. Water Resources Center Report No. 84. Davis, CA: University of California Centers for Water and Wildland Resources, pp. 35–40.

King, Gary, Robert Keohane, and Sidney Verba (1994) *Designing Social Inquiry: Scientific Inference in Qualitative Research, Princeton,* NJ: Princeton University Press.

Knapp, Keith C. and Henry J. Vaux (1982) Barriers to Effective Ground-Water Management: The California Case. *Ground Water* 20(1): 61–66.

Loehman, Edna Tusak and Ariel Dinar (1995) Introduction. In *Water Quantity/Quality Management and Conflict Resolution: Institutions, Processes, and Economic Analyses.* Ariel Dinar and Edna Tusak Loehman, eds. Westport, CT: Praeger, pp. xxi–xxx.

Loehman, Edna Tusak and D. Marc Kilgour (1998) Introduction: Social Design for Environmental and Resource Management. In *Designing Institutions for Environmental and Resource Management.* Edna Tusak Loehman and D. Marc Kilgour, eds. Cheltenham, UK: Edward Elgar, pp. 1–25.

Mallery, Michael (1983) Groundwater: A Call for a Comprehensive Management Program *Pacific Law Journal* 14(4): 1279–1307.

McClurg, Sue (1996) Maximizing Groundwater Supplies. *Western Water* (May/June): 4–13.

McCoy, Patricia R. (2003) Idahoans Debate Merits of Threatened Water Suit (Boise, Idaho) *Capital Press Agriculture Weekly* (September 16).

Metropolitan Water District of Southern California (1986) High-Flowing Colorado River Roars into the Record Books. *MWD Focus* 1986(4): 7.

——(1993) MWD, Semitropic Agree on Water Banking. *MWD Focus* 1993(2): 7.

——(1996a) *Southern California's Integrated Water Resources Plan.* Report No. 1107. Volume 1: The Long-Term Resources Plan. Los Angeles, CA: Metropolitan Water District of Southern California.

——(1996b) *Southern California's Integrated Water Resources Plan.* Report No. 1107. Volume 2: Metropolitan's System Overview. Los Angeles, CA: Metropolitan Water District of Southern California.

Metzler, Frank and Tom Carr (1998) CAP Supply for the 21st Century: Taking Our Droughts to the Bank. In *Water at the Confluence of Science, Law, and Public Policy: Proceeding of the Eleventh Annual Symposium, Arizona Hydrological Society,* September 24–26, Tucson, AZ.

Moore, Deborah and Zach Willey (1991) Water in the American West: Institutional Evolution and Environmental Restoration in the 21st Century. *University of Colorado Law Review* 62(4): 775–825.

National Groundwater Policy Forum (1986) *Groundwater: Saving the Unseen Resource.* Washington, DC: The Conservation Foundation.

National Water Commission (1973) *Water Policies for the Future: Final Report to the President and to the Congress of the United States.* Port Washington, NY: Water Information Center, Inc.

Noel, J.E., B. Delworth Gardner, and Charles V. Moore (1980) Optimal Regional Conjunctive Water Management. *American Journal of Agricultural Economics* 62(3): 489–498.

Ostrom, Elinor (1986) An Agenda for the Study of Institutions. *Public Choice* 48: 3–25.

——(1998) The Institutional Analysis and Design Approach. In *Designing Institutions for Environmental and Resource Management.* Edna Tusak Loehman and D. Marc Kilgour, eds. Cheltenham, UK: Edward Elgar, pp. 68–90.

——(1999) Institutional Rational Choice: An Assessment of the Institutional Analysis and Development Framework. In *Theories of the Policy Process.* Paul Sabatier, ed. Boulder, CO: Westview Press.

Ostrom, Elinor, Roy Gardner, and James Walker (1994). *Rules, Games, and Common–Pool Resources.* Ann Arbor, MI: University of Michigan Press.

Ostrom, Vincent (1953) *Water and Politics: A Study of Water Policies and Administration in the Development of Los Angeles.* Los Angeles, CA: Haynes Foundation.

Pielke, Roger A. Sr., Nolan J. Doesken, and Jose D. Salas (2001) Drought Threat to Colorado Water. *Colorado Water: Newsletter of the Water Center at Colorado State University* 18(4): 7–11.

Reiter, Stanley (1998) On Coordination, Externalities, and Organization. In *Designing Institutions for Environmental and Resource Management.* Edna Tusak Loehman and D. Marc Kilgour, eds. Cheltenham, UK: Edward Elgar, pp. 57–67.

Ruttan, Vernon W. (1998) Designing Institutions for Sustainability. In *Designing Institutions for Environmental and Resource Management.* Edna Tusak Loehman and D. Marc Kilgour, eds. Cheltenham, UK: Edward Elgar, pp. 142–161.

Sample, Kathy, Bijou Irrigation Company, interview in Fort Morgan, Colorado, October 1998.

Santa Fe New Mexican (2002) County to Take Close Look at Injection Idea. *The Santa Fe New Mexican* (August 4).

Schneider, Anne J. (1994) Are Our Ground Water Laws Adequate? *Proceedings, Nineteenth Biennial Conference on Ground Water.* Water Resources Center Report No. 84. Davis, CA: University of California Centers for Water and Wildland Resources, pp. 47–54.

Sherow, James E. (1989) The Chimerical Vision: Michael Creed Hinderlider and Progressive Engineering in Colorado. *Essays in Colorado History* No. 9: 37–59.

Silvo, Andrew (2002) County to Consider Aquifer Protection. *San Bernardino County Sun* (October 9).

Smith, Zachary A. (1984) Rewriting California Groundwater Law: Past Attempts and Prerequisites to Reform. *California Western Law Review* 20(2): 223–257.

———(1989) *Groundwater in the West.* San Diego, CA: Academic Press.

Soltis, Dan (1997) Aquifer Recharge Experiment Termed "Proven Success." *Water Engineering & Management* 141(8): 8.

Southwest Hydrology (2003) Nevada Requests More Water. *Southwest Hydrology* 2(5): 11.

Sullivan, Jim (1998) Water for All: Planning for Region a Challenge. *The Denver Post* (January 18), p. E–1.

Sutherland, P. Lorenz and John A. Knapp (1988) The Impacts of Limited Water: A Colorado Case Study. *Journal of Soil and Water Conservation* 43(4): 294–298.

Tarlock, A. Dan, James Corbridge, and David Getches (1993) *Water Resource Management: A Casebook in Law and Public Policy,* 4th edit. Westbury, NY: The Foundation Press.

Taylor, Michael and Sara Singleton (1993) The Communal Resource: Transaction Costs and the Solution of Collective Action Problems. *Politics and Society* 21(2):195–214.

Tellman, Barbara (1996) Why Has Integrated Management Succeeded in Some States But Not in Others? *Water Resources Update* No. 106: 13–18.

Thomas, Clive E., ed. (1991) *Politics and Public Policy in the Contemporary American West.* Albuquerque, NM: University of New Mexico Press.

Thomas, Gregory A., David K. Fullerton, David R. Purkey, Lee Axelrod, Timothy Ramirez, and Marcus Moench (1997) Feasibility Study of a Maximal Groundwater Banking Program for California: A Working Draft. Unpublished manuscript, 67 pp.

Thomas, R.O. (1955) General Aspects of Planned Ground-Water Utilization. *Proceedings, American Society of Civil Engineers* Volume 81, Item 706, 11 pp.

Thorson, John E. (1994) Protecting Ground Water as *Res Publica. Proceedings, Nineteenth Biennial Conference on Ground Water.* Water Resources Center Report No. 84. Davis, CA: University of California Centers for Water and Wildland Resources, pp. 1–7.

Thorson, Norman W. (1978) Storing Water Underground: What's the Aqui–fer? *Nebraska Law Review* 57(3): 581–632.

Trager, Susan M. (1988) Emerging Forums for Groundwater Dispute Resolution in California: A Glimpse at the Second Generation of Groundwater Issues and How Agencies Work Towards Problem Resolution. *Pacific Law Journal* 20(1): 31–74.

——— (1994) When Is a Law to Protect Water Quality or an Endangered Species Actually a Scheme of the Federal Government to Regulate Ground Water? *Proceedings, Nineteenth Biennial Conference on Ground Water.* Water Resources Center Report No. 84. Davis, CA: University of California Centers for Water and Wildland Resources, pp. 143–148.

Trelease, Frank J. (1982) Conjunctive Use of Groundwater and Surface Water. *Rocky Mountain Mineral Law Institute Journal* 27B: 1853+

U.S. Fish and Wildlife Service (1997) *Cooperative Agreement for Platte River Research and Other Efforts Relating to Endangered Species Habitat Along the Central Platte River, Nebraska* (July).

Veblen, Thomas T. and Diane C. Lorenz (1991) *The Colorado Front Range: A Century of Ecological Change.* Salt Lake City, UT: University of Utah Press.

Weatherford, Gary D. (1994) Institutional Dimensions of Sustainability. *Proceedings, Nineteenth Biennial Conference on Ground Water.* Water Resources Center Report No. 84. Davis, CA: University of California Centers for Water and Wildland Resources, pp. 17–23.

Williamson, Oliver (1985) *The Economic Institutions of Capitalism.* New York: Free Press.

Index